A POLITICAL CHRONOLOGY OF THE AMERICAS

A POLITICAL CHRONOLOGY OF THE AMERICAS

FIRST EDITION

EUROPA PUBLICATIONS
Taylor & Francis Group

First Edition 2001

11 New Fetter Lane, London EC4P 4EE, United Kingdom
(A member of the Taylor & Francis Group)

ISBN 1-85743-118-9

Editors: David Lea, Colette Milward
Assistant Editor: Annamarie Rowe
Special Editorial Assistance: Lucy Dean, Driss Fatih, Iain Frame, Anthony Gladman, Tiara Misquitta, Katharine Murison, Alison Neale, Daniel Ward, Jacqueline West
Contributors: Owain Johnson, Sondeep Kandola, Dr Ana Vadillo, Linda Van Buren

Data manipulation and database design: Mark Wilson, Bibliocraft Ltd

Typeset by Bibliocraft Ltd, Dundee
Printed and bound by TJ International Limited, Trecerus Industrial Estate, Padstow, Cornwall

Foreword

This is the last title in a new six-volume series of Political Chronologies of the World. Each of the other volumes in the series concerned a different region of the world: respectively, Europe, Central, South and East Asia, and the Middle East, Africa and South-East Asia and Oceania.

Although the book includes greater coverage of more-recent events—particularly in countries with a recent history of political upheaval—it also provides invaluable detail on the early history of each nation. Each chronology begins at least as early as the emergence of an entity resembling the modern nation, and in many cases considerably earlier. The prominence of the USA in international affairs necessitates that only its most significant interventions in this sphere appear.

In addition to coverage of purely political events, each title in the series also includes details of the principal economic, cultural and social landmarks in the history of each nation. Examples might include the adoption of new languages, alphabets, calendars or religions.

This series aims to be the first point of reference for concise information on the history of each nation in the world. It is hoped that the volumes in the series will enable readers easily to locate self-contained entries on the period and area in which their interest lies.

November 2001

Contents

Abbreviations

AD	*anno Domini*
Adm.	Admiral
a.m.	*ante meridiem* (before noon)
BC	before Christ
Brig.	Brigadier
c.	*circa*
Cdre	Commodore
CIS	Commonwealth of Independent States
Cmdr	Commander
Col	Colonel
Dr	Doctor
EEC	European Economic Community
EU	European Union
etc.	et cetera
f.	founded
Gen.	General
HM	His (or Her) Majesty
kg	kilogram(s)
km	kilometre(s)
kW	kilowatt(s)
Lt	Lieutenant
m.	million
Maj.	Major
Mgr	Monsignor
MP	Member of Parliament
NATO	North Atlantic Treaty Organization
OAS	Organization of American States
Prof.	Professor
Rev.	Reverend
St	Saint
UN	United Nations
UNHCR	United Nations High Commissioner for Refugees
US	United States
USA	United States of America
USS	United States Ship
USSR	Union of Soviet Socialist Republics

Antigua and Barbuda

***c.* 2400 BC:** Antigua was first settled by the Siboney ('stone people' in Arawak), who were Meso-Indians from South America.

AD 35–1100: Following the departure of the Siboney, Antigua was inhabited by the Arawaks.

1100: Carib Indians displaced the Arawaks.

1493: The Italian navigator in the service of the Spanish Crown, Cristoforo Colombo (Christopher Columbus) passed Antigua during his second Caribbean voyage. He named the island after Santa María la Antigua, the patron saint of Seville, Spain.

1630s: The British brought the first African slaves to Antigua.

1632: A group of Englishmen from the nearby island of Saint Christopher (St Kitts) established the first settlement on Antigua.

1661: Settlers from Antigua colonized the island of Barbuda.

1666: French troops landed at and occupied Antigua for eight months.

1667: France formally renounced its claim to Antigua at the Treaty of Breda and withdrew its troops from the island.

1684: Sir Christopher Codrington began large-scale sugar cultivation in Antigua. The plantation system flourished, and by the 18th century the island supported more than 150 large plantations.

1685: Codrington leased Barbuda from the British Crown.

1834: The United Kingdom abolished slavery throughout its empire. Antigua immediately emancipated all slaves, unlike other colonies, which introduced a four-year waiting period.

1850s: The sugar industry dramatically declined in importance.

1860: The island of Barbuda, formerly the private property of the Codrington family, reverted to the British Crown and was annexed to Antigua.

1871: Antigua and Barbuda began to be administered together as part of the Leeward Islands Federation, which also included St Kitts and Nevis, Anguilla, Montserrat, the British Virgin Islands and (until 1940) Dominica.

1939: The country's first labour movement, the Antigua Trades and Labour Union, was founded.

1946: The president of the Antigua Trades and Labour Union, Vere Cornwall Bird, founded the island's first political party, the Antigua Labour Party (ALP).

9 February 1948: Earl Baldwin of Bewdley was appointed Governor of the Leeward Islands.

11 May 1950: Earl Baldwin resigned as Governor and was replaced by K. W. Blackburne.

1956: The ALP won all the eight elected seats in the Legislative Council and Bird was appointed Chief Minister.

30 June 1956: The Leeward Islands Federation was dissolved. Antigua became a separate colony and prepared to enter the proposed British Caribbean Federation as an independent unit.

2 August 1956: The British Parliament adopted a bill establishing the British Caribbean Federation, which was to have a joint Government, Legislature and Supreme Court.

October 1956: Alexander Williams replaced Blackburne as Governor of the Leeward Islands.

January 1958: Antigua joined the West Indies Federation, which included almost all of the United Kingdom's Caribbean colonies and aimed to move towards independent statehood.

6 October 1959: The British Colonial Office announced the abolition of the post of Governor of the Leeward Islands as part of the reform of the islands' constitutions. The reforms were designed to reduce the administrative differences between the members of the West Indies Federation.

28 November 1961: At elections to the expanded Legislative Council, the ALP won all 10 elected seats and Bird was reappointed Chief Minister.

27 February–5 March 1962: Representatives of Barbados and the seven Leeward and Windward Island colonies met in Barbados and requested that the British Government approve a new federation, known as the Little Eight, in the event of the collapse of the West Indies Federation.

31 May 1962: The West Indies Federation was formally dissolved.

27 February 1967: Antigua and Barbuda's status changed from that of a colony to an Associated State. The islands were granted full internal self-government, while the United Kingdom retained responsibility for defence and foreign relations. Antigua became an associated state of the Commonwealth, with Barbuda and the uninhabited island of Redonda as dependencies. A House of Representatives replaced the Legislative Council, the Administrator became the Governor and the Chief Minister was restyled Premier.

February 1971: The Progressive Labour Movement (PLM) decisively beat the ALP at legislative elections, and the leader of the PLM, George Walter, became Premier.

1972: Sugar production ceased, after 400 years as the islands' main industry. Subsequent attempts to revive the industry failed.

18 February 1976: The ALP returned to power after winning 11 of the 17 seats in the House of Representatives and Bird became Premier.

24 April 1980: The ALP was re-elected with an increased majority. Bird had called the elections almost a year early in order to establish a clear mandate for talks on independence with the British Government.

1 November 1981: The islands achieved full independent status under the name Antigua and Barbuda. Executive power was vested in the Governor-General, representing the British crown, and in Bird, who became Prime Minister.

17 April 1984: The ruling ALP won an overwhelming victory at legislative elections, and Bird retained his position as Prime Minister.

1 January 1987: Bird carried out a cabinet reshuffle in which he replaced those ministers who had urged him to dismiss his elder son, Vere Bird, Jr, the Minister of Public Utilities, who had been implicated in a financial scandal.

9 March 1989: The ALP retained 15 of its 16 seats at legislative elections and Bird was returned to office for the fourth consecutive time.

15 November 1990: Vere Bird, Jr, was dismissed from the Cabinet and banned for life from holding office in the Government, following the discovery by a commission of inquiry that he had ordered arms from an Israeli manufacturer that were later supplied to a Colombian drugs cartel.

15 March 1991: Bird effected a reshuffle of the Cabinet, in an attempt to increase his own responsibilities. Several ministers, including his younger son Lester Bird, criticized this decision and refused to join the new Cabinet.

3 September 1991: The rebel ministers rejoined the Cabinet.

February 1992: A number of violent incidents were reported following public demonstrations against alleged corruption in Bird's Government.

10 June 1993: James (later Sir James) Carlisle replaced Sir Wilfred Jacobs as Governor-General.

February 1994: Bird resigned as Prime Minister and was replaced by his son and the new leader of the ALP, Lester Bird.

8 March 1994: The ALP secured a fifth consecutive term in office, despite the serious corruption allegations facing the party's leadership, and Lester Bird was reappointed Prime Minister.

14 May 1996: Bird announced a major cabinet reshuffle, which controversially included the appointment of his disgraced older brother, Vere Bird, Jr, as Special Adviser on Telecommunications, Science, Technology and Information.

22 September 1996: The Minister of Finance and Social Security, Molwyn Joseph, resigned following revelations that he had evaded customs duties.

1997: The Government announced the closure of 11 offshore banks as part of its attempts to eradicate incidences of money-laundering on the island.

3 December 1997: The disgraced former Minister of Finance, Molwyn Joseph, was reappointed to the Cabinet as Minister of Planning and Implementation.

16 December 1997: Vere Bird, Jr, was wounded in a shooting incident on the same day as the Government agreed terms for the compulsory resettlement of Cyril 'Taffy' Bufton, the sole inhabitant of Guiana Island. Bufton was charged with attempted murder the following year.

1998: The Government closed down six Russian-owned banks accused of money-laundering and approved new legislation to regulate the financial sector; however, a report published the following year by the US Government suggested that the new legislation was inadequate.

9 March 1999: The ALP secured its sixth successive general-election victory and Bird was reappointed Prime Minister. The ALP won 12 seats in the 17-seat legislature, the United Progressive Party (UPP) four and the Barbuda People's Movement one. The new cabinet included the Prime Minister's brother, Vere Bird, Jr, as Minister of Agriculture, Lands and Fisheries.

September 1999: The Government established an independent body to regulate the country's 'offshore' banking services that had previously attracted international criticism, and pledged to introduce more stringent controls.

June 2000: The Organisation for Economic Co-operation and Development (OECD) and the Financial Action Task Force (FATF, part of the OECD) criticized Antigua and Barbuda's financial system. The OECD urged the Government to reform its rules by 31 December 2005 or face penalties.

February 2001: The head of the Barbuda Council, Arthur Nibbs, accused Bird of using excessive power to dominate Barbuda.

22 May 2001: Bird dismissed the Minister of Health, Bernard Percival, and the Attorney-General, Errol Cort, following their implication in irregular administrative procedures.

Argentina

c. **AD 500:** The southern areas of the continent were inhabited by various tribal peoples. The largest and apparently most advanced group were the Diaguita, who had established a complex society based on the cultivation of maize in the present-day Calchaquí valley. The Comechingone confederation also existed in the area of what is now Córdoba and the Guaraní hunter-gather tribes in Chaco and Patagonia.

late 15th century: The Diaguita began to establish political links with the Inca Empire, based in what is now Peru; by the middle of the next century, the religion, language and art of the Diaguita displayed Incan influences.

1516: A Spanish navigator, Juan Díaz de Solis, discovered the estuary subsequently known as the Río de la Plata, while on an expedition to find the south-west passage to the East Indies. He claimed the area for the Spanish crown of Castile.

1526: A more detailed reconnaissance of the area was conducted for the Spanish crown by the Italian-born explorer, Sebastiano Caboto. A shortage of supplies forced the expedition to move further up the River Paraná, were the fort Sancti Espiritus was founded near present-day Rosario.

1531: The Portuguese crown dispatched its own navigator, Afonso de Souza, to the Plata region, who reported an abundance of gold and silver.

1535: Pedro de Mendoza was appointed military governor of the Plata by King Carlos V of Castile and commissioned to establish the region's first settlement, which was to be named Puerto Nuestra Señora del Buen Aire. The colonization project was troubled by shortages of food and the hostility of local indigenous groups.

1537: A more successful settlement was founded by Mendoza's troops further up the Paraná River, which was named Asunción. The land was more fertile and relations with the local Guaraní people peaceful. Domingo de Irala was appointed Governor of the settlement, a post which he held for the ensuing 30 years.

1541: The original Spanish settlement of Buen Aire was abandoned.

1553: Francisco de Aguirre and a group of settlers from Peru established the settlement of Santiago del Estero. Several other settlements were also founded in the area.

1562: Santiago del Estero was the only settlement to survive a spate of attacks by indigenous groups in the northern regions.

1573: Juan de Garay founded Santa Fe. Roman Catholic missionaries, the largest group of whom were Jesuits, also established settlements.

1580: The re-colonization of Buenos Aires began, supported by supplies of livestock, grain and weapons from Santa Fe and Asunción. The Spanish attempted to impose a programme of enforced labour on the indigenous tribes, to assist with construction of the towns and the cultivation of crops.

1610: The administration of the larger settlements was conducted by officials called *cabildos*; the offices were sold to colonizers, who became responsible for food supplies and the subjugation of the local indigenous people.

1620: The La Plata region was incorporated into the Viceroyalty of Peru.

1622–23: The Spanish Government, in an attempt to maintain a monopoly of trade in Buenos Aires, introduced a tariff line, the *Aduana Seca*, at Córdoba and also attempted to prevent the use of money in the town. The authorities were concerned that the large numbers of Portuguese merchants there would export silver to Spain's European enemies.

late 17th century: A trade in contraband flourished in Buenos Aires; the rest of the colony suffered from a decline in demand for South American silver.

1702: King Felipe V of Spain authorized the French Guinea Company to export slaves to Spanish South America through Buenos Aires; this encouraged further contraventions of the Government's monopoly. This concession was transferred to the British South Sea Company in 1714.

1739: The British were expelled from Buenos Aires, following a war between Spain and the United Kingdom in Europe.

1776: The area occupied by what is now Argentina, Bolivia and Uruguay was separated from Peru and administered as the Viceroyalty of Río de la Plata. The Spanish Government also subsidized the development of Buenos Aires as a strategic military base.

June 1806: Buenos Aires was attacked by a British naval fleet. The Viceroy fled to Córdoba, leaving the citizens to form militias to repel the invaders.

1807: The citizen militias successfully defeated a further invasion attempt by British naval forces; this gave rise to the beginnings of an independence movement.

1808: Following the deposition of Fernando VII of Spain by Napoléon Bonaparte I of France, the militias of Buenos Aires refused to recognize the appointment of Bonaparte's brother, Joseph, as Spanish monarch. Martín de Algaza attempted a *coup d'état* against the new viceroy.

28 May 1810: The citizen militias of Buenos Aires overthrew the viceroy and established their own administration in the name of Fernando VII. They subsequently launched a campaign to liberate the rest of La Plata.

1812–13: The separatists gained some victories over the armies of Joseph Bonaparte; the liberated areas, which included Corrientes, Rios and Santa Fe, were divided into 14 provinces for administrative purposes.

1813: The northern provinces seceded from the Viceroyalty as Paraguay, with its capital at Asunción.

1814: Under the command of José de San Martín, the rebel armies won a number of decisive campaigns against royalist armies in Chile and Peru.

9 July 1816: Following an earlier meeting of representatives from the liberated provinces, independence from Spanish rule was declared and the United Provinces of South America—subsequently the United Provinces of Río de la Plata–was established. A 'supreme director' was also appointed as head of state.

1819–20: A dispute between the Unitarians of Buenos Aires, who favoured a highly centralized form of government, and the Federalists of the provinces, who preferred a federal system of government, precipitated civil war. Although hostilities ceased in 1820, a permanent system of government was not established.

1821: Brazil annexed Uruguay.

1825: Bolivia seceded from the United Provinces.

1825–28: The United Provinces and Brazil entered into a war for sovereignty over Uruguay, which emerged from the conflict as an independent state.

1829: A federalist, Gen. Juan Manuel de Rosas, was elected Governor of Buenos Aires. He succeeded in uniting the provinces into an Argentine Confederation by 1832.

1852: Gen. Justo Urquiza, the former Governor of Uruguay and Brazil, led a *coup d'état* against Rosas. The following year, a Constitution was established and Urquiza was appointed President of the Republic of Argentina; the authorities of Buenos Aires objected the terms of the Constitution and failed to recognize the republic.

1853: The provincial Government of Corrientes instituted a programme which encouraged settlers from France to establish farms in the region. Other provincial administrations organized similar programmes, attracting immigrants from Italy, Germany and Switzerland.

1859: Buenos Aires rebelled against the republic; following the defeat of its forces in October, the city was compelled to recognize the Constitution.

1861: A further rebellion by Bartolomé Mitre in Buenos Aires was successful and Urquiza was forced to resign.

May 1862: Mitre was elected as President of the Republic and Buenos Aires' position as the national capital was affirmed.

1865: The Welsh colony of Chubut was founded in Patagonia.

1865–70: Paraguay invaded the north-eastern regions; the Government formed an alliance with Brazil and Uruguay, known as the Triple Alliance, which eventually defeated the invaders.

1874: Nicolas Avellaneda was elected President; a rebellion by supporters of Mitre was subsequently suppressed.

1879–80: Gen. Julio Roca led a campaign to subjugate the indigenous tribes of the Pampas region. Following his success, he was elected President.

1881: In an agreement with Chile, the Government acquired the eastern half of the Tierra del Fuego.

1895: Argentina acquired land from Brazil, following the resolution of a boundary dispute.

1899–1902: The United Kingdom mediated in a dispute between the Government and Chile regarding the Patagonian border.

1912: The administration of Roque Sáenz Peña introduced legislation which provided for elections to be conducted by secret ballot, in an attempt to reduce electoral corruption.

1914: The Government mediated in a frontier dispute between the USA and Mexico.

1914–18: Argentina supplied the Allied Powers with foodstuffs and raw materials during the First World War.

1916: The leader of the Unión Civica Radical (UCR), Hipólita Yrigoyen Alem, was elected to a six-year term as President in the first elections held under universal male suffrage.

October 1922: Marcelo de Alvear Pacheco of the UCR was elected President in succession to Yrigoyen.

October 1928: Yrigoyen was elected to a second presidential term.

September 1930: Yrigoyen's Government was deposed in a military *coup d'état* led by José Uriburu y Uriburu.

February 1932: Following Uriburu's death, Agustín Justo Rolón became President.

1933–34: The Radicals instigated a rebellion against the Government and civilian rule was re-established.

1936: Right-wing fascist movements united to form the Frente Nacional, which advocated the establishment of a personalist government.

1937: The Frente Nacional candidate, Roberto Ortiz Lizardi, was elected President.

1939: The Government declared the country's neutrality on the outbreak of the Second World War. The war created divisions within Argentine society, with elements supporting both the Allies and the Axis powers.

July 1940: President Ortiz fell ill and appointed the Vice-President, Ramón Castillo Barrionuevo, to perform his functions in a temporary capacity.

June 1942: Ortiz resigned and Castillo formally became President.

1943: The pro-British military leader, Gen. Arturo Rawson Corvalán, led a military *coup d'état* against President Castillo. Gen. Pedro Ramírez Machuca was subsequently installed to lead a military Government; his administration abolished all political parties and closed all the opposition newspapers.

January 1944: The Government terminated diplomatic relations with Japan and Germany.

February 1944: A military junta, led by Col Juan Domingo Perón Sosa, deposed Ramírez's successor, Edelmiro Farrell.

March 1945: The Government declared war on Germany and Japan; the following month, it signed the Act of Chapultec, by which it agreed to assist fellow American nations in the event of aggression by a third party.

June 1945: Argentina became a founder member of the UN.

24 February 1946: Perón won presidential elections following an aggressive campaign by the Partido Justicialista (PJ–later known as the Peronists), which gained most support among the agricultural and industrial working classes. Perón's wife, Eva Duarte de Perón, subsequently controlled, in an unofficial capacity, the Government's policy on labour relations and social services.

July 1949: The PJ renominated Perón as its candidate for presidential elections scheduled to take place in 1952. This prompted much criticism among opposition parties, causing the Government to introduce legislation which extended prison terms for dissidents and provided for the suppression of the media. Perón subsequently succeeded in amending the Constitution to permit the President to serve more than one term of office.

February 1951: Perón was re-elected President, the PJ gaining 135 of 149 seats in the House of Deputies. Perón's rule became increasingly personal, although his wife's popularity with the country's population ensured that disorder was minimal.

1954: Following his accusation that Roman Catholic clergyman had encouraged anti-Government sentiment, Perón introduced legislation which legalized divorce, bestowed all the benefits of legitimacy to children born out of wedlock and legalized prostitution.

16 June 1955: A rebellion by dissident elements of the military was suppressed in Buenos Aires.

September 1955: A further rebellion escalated into a brief period of civil war in which the Government was removed from office and Maj.-Gen. Eduardo Lonardi Doucet installed as provisional President. Perón was exiled, first to Paraguay and later to Spain.

November 1955: Lonardi was forced to resign by a further military intervention and Maj.-Gen. Pedro Eugenio Aramburu Cilveti was installed as President; his administration restored the Constitution of 1853, preventing the head of state from seeking re-election.

June 1956: Supporters of Perón led a revolt which was severely suppressed by the armed forces; 38 Peronists were executed, and several thousand more arrested and imprisoned.

July 1956: Elections to the Constituent Assembly took place (the Peronists were excluded from participation); the UCR gained the most votes, followed by the Unión Cívica Radical Intransigente (UCRI).

1958: The leader of the UCRI, Arturo Frondizi Ercolí, won presidential elections with the support of Peronist and Communist factions.

1960: The Government became a member of the Latin American Free Trade Association, which aimed to develop trading relations with other countries in the region.

March 1962: In legislative elections, the Peronists gained 35% of votes cast. President Frondizi was subsequently accused of sympathies with the Peronist movement and deposed by the military; José María Guido was appointed his successor.

July 1963: In new elections, from which the Peronists and Communists were banned, Arturo Illía Francesconi of the Unión Cívica Radical del Pueblo (UCRP), was elected President.

June 1966: President Illía was deposed in a *coup d'état*; a military junta, led by Gen. Juan Carlos Onganía Carballo, subsequently took control.

June 1970: Gen. Onganía was deposed by other members of the military junta; Roberto Levingston Laborda assuming the leadership.

March 1971: Gen. Alejandro Agustín Lanusse Gelly assumed the leadership of the military junta.

1972: Price rises and agitation by Peronistas caused civil unrest, precipitating the declaration of states of emergency in many of the provinces during the year. Gen. Lanusse announced that legislative elections, in which Peronist groups (although not Perón himself) would be permitted to participate, would be held in March 1973.

March 1973: The Frente Justicialista de Liberación (FJL), which was supported by the Peronists, won legislative elections; their candidate, Hector Cámpora Demaestre, was subsequently elected President.

20 June 1973: Perón returned to Buenos Aires; his return precipitated rioting in which some 380 people died.

12 July 1973: Amid increasing unrest among supporters and opponents of Perón, and between extremist elements within the Peronist movement, Campora resigned; new elections were organized for September.

September 1973: Perón was elected President, gaining 61% of votes cast; he appointed his third wife, María Estela (Isabelita) Martínez de Perón, as Vice-President.

1 July 1974: Perón died; he was succeeded by 'Isabelita' de Perón. Her Government's programme of economic austerity prompted a revival of civil unrest and strikes, and failed to prevent the deterioration of the country's economy.

December 1975: Units of the air force attempted to rebel against the Government.

24 March 1976: Isabelita Perón was deposed by a military junta, led by Lt-Gen. Jorge Rafael Videla Redondo. The junta dissolved the legislature, suspended all political-party and trade-union activity and imposed martial law.

1977: The Argentine Commission for Human Rights alleged that the military regime was responsible for 2,300 political murders, 10,000 political arrests and some 30,000 'disappearances'.

March 1981: Videla (who had left his military position) was succeeded by Gen. Roberto Viola Prevendini, who announced his intention that democratic rule be restored.

December 1981: Owing to ill health, Viola resigned; he was replaced by the Commander-in-Chief of the armed forces, Gen. Leopoldo Galtieri. The regime's suppression of opposition continued and, in some cases, intensified.

2 April 1982: Gen. Galtieri ordered the occupation of the British-held Falkland Islands (Islas Malvinas—an archipelago in the South Atlantic, sovereignty over which had been disputed since the United Kingdom claimed the islands in 1833); British forces were dispatched to recapture the islands.

14 June 1982: The Falkland Islands were recaptured by the British; Gen. Galtieri was forced to resign and Maj.-Gen. Reynaldo Bigone Ramayón was appointed head of state.

30 October 1983: General elections were held. The UCR won 317 of the 600 available seats in the presidential electoral college and 129 of the 254 seats in the Chamber of Deputies. The UCR's candidate, Raúl Alfonsín Foulkes, a lawyer who had opposed the war and was active in defence of human rights during the military regime, was elected President.

December 1983: Following the discovery of mass graves, the Government announced the establishment of a National Commission of Disappearance of Persons to investigate the activities of the successive military juntas, which became known as the *guerra sucia* ('dirty war'). President Alfonsín also ordered the court-martial of the members of the first three military juntas to rule following the 1976 *coup d'état*.

April 1985: The trials of the former military leaders commenced; witnesses testified to systematic atrocities and accused the former regimes of instigating campaigns of terror.

May 1985: The Government concluded an agreement with Chile, regarding sovereignty over the Beagle Islands; it was decided that while Chile would have sovereignty over the islands, Argentina would be able to exploit petroleum and other mineral resources present in the surrounding waters.

15 June 1985: The peso argentino was replaced as the country's currency by the austral, in an attempt to control inflation (one austral was equivalent to 1,000 pesos argentinos).

December 1985: Videla and Adm. Eduardo Massera were sentenced to life imprisonment for violations of human rights committed during the country's period of military rule; five others received shorter sentences.

May 1986: Three members of the junta which held power during the Falklands-Malvinas conflict, were convicted of negligence and received prison sentences; Gen. Galtieri was sentenced to 12 year's imprisonment.

December 1986: The Government approved the *Punto Final* ('Full Stop') Law, which authorized civil and military courts to begin the trials of members of the armed forces who had been accused of violations of human rights, within a 60-day period, ending on 22 February 1987.

May 1987: Legislation providing for an amnesty for all but the senior ranks of the police and the armed forces was introduced.

September 1987: In elections to the Chamber of Deputies, the UCR lost 13 of its 130 seats, thus losing its majority; The Movimiento Nacionalista Justicialista–Peronist increased its number of seats from 101 to 105.

January 1988: Lt-Col Aldo Rico led a rebellion by the armed forces, who demanded an increase in salaries and an amnesty for officers awaiting trial for human-rights violations. A similar revolt was instigated by Col Mohamed Ali Seineldín in December.

14 May 1989: Carlos Saúl Menem Akim, the candidate of the Frente Justicialista de Unidad Popular (FREJUPO), an electoral alliance formed by the Partido Democrática Cristiano (PDC) and the Partido Intransigente (PI) won 48.5% of the votes cast in the presidential election; the alliance was also the largest party in the Chamber of Deputies, winning 66 of the 127 seats.

October 1989: The new Government issued a pardon to 210 military officers and soldiers who had been involved in the 'dirty war'.

December 1989: The Minister of the Economy, Nestor Rapanelli, resigned following his failure to prevent hyperinflation.

February 1990: Full diplomatic relations were restored with the United Kingdom.

December 1990: The release of Videla, Gen. Viola and Adm. Massera by the Government precipitated mass demonstrations in Buenos Aires.

January 1991: In an attempt to control inflation and stabilize the economy, the austral, which had experienced a number of devaluations, was fixed at 10,000 to the US dollar, supported by the Central Bank's reserves.

26 March 1991: The Treaty of Asunción was signed by Argentina, Brazil, Paraguay and Uruguay, committing the four countries to the establishment of a common market, to be known in Portuguese as the Mercado Comum do Sul and in Spanish as the Mercado Común del Sur (Mercosul/Mercosur–the Southern Common Market).

October 1991: In an attempt to reduce hyperinflation, the President decreed that all state regulation of the economy should cease.

1 January 1992: The austral was replaced as the country's currency by the nuevo peso argentino, equal to US $1.

November 1992: The country's largest trade-union organization, the Confederación General del Trabajo (CGT) announced a one-day general strike.

December 1993: The UCR and the Peronist Partido Jusiticialista (PJ) concluded an electoral-reform agreement, which provided for the amendments to the Constitution, including the re-election of a president for a consecutive term, the reduction of the term of office to four years and the abolition of the presidential electoral college.

April 1994: In elections to the 305-seat Constituent Assembly, the PJ won 37.7% of the votes cast and 136 seats, the UCR securing only 19.9% of the ballot.

24 August 1994: A new Constitution, which included clauses on the protection of human rights, was promulgated. Under the provisions of the new document, the president was thenceforth to be directly elected, with a second round of voting required if the most popular candidate received less than 45% of the votes cast, but not if that candidate received 40% of the ballot and the nearest challenger less than 30%. Other provisions included the delegation of some presidential powers to a Chief of Cabinet and the reiteration of the territorial claim to the Falkland Islands/Malvinas.

31 December 1994: Mercosul/Mercosur was established; its provisions came into effect the following day.

14 May 1995: Menem was re-elected to the presidency, receiving 49.9% of the votes cast. An alliance of left-wing parties, the Frente del País Solidario (Frepaso) won the most (35%) votes in the Chamber of Deputies. Menem formed a new Cabinet (with few changes from the previous administration)–Eduardo Bauzá was appointed to the post of Chief of Cabinet.

March 1996: Bauzá resigned and was succeeded by Jorge Rodríguez, hitherto the Minister of Education and Culture.

April 1996: The value of the nuevo peso against the US dollar was fixed at US $1 = 0.9995 nuevos pesos.

June 1996: In elections for the office of Mayor and to the constituent assembly of Buenos Aires (the offices of the autonomous government having been established in the Constitution of August 1994), the UCR made considerable gains.

July 1996: The Minister of Justice, Rodolfo Barra, and the Minister of Defence, Oscar Camilión, resigned, following accusations of corruption.

26 July 1996: Domingo Cavallo, the reformist economy minister, was dismissed; he was replaced by Roque Fernández, hitherto the President of the Central Bank.

December 1996: In an attempt to increase the pace of the reform of labour legislation, Menem attempted to amend the laws using presidential decree; the reforms were declared unconstitutional by the Supreme Court the following month.

March 1997: Cavallo founded a new political party, the Acción por la República (AR) and subsequently formed an alliance with the Nueva Dirigencia (ND).

May–August 1997: Widespread social and industrial unrest occurred.

26 October 1997: In a congressional election, an alliance of Frepaso and UCR won 61 of the 127 seats, the PJ winning 51 seats and the AR-ND three.

June 1998: Videla was arrested in connection with the abduction and illegal adoption of children whose parents had 'disappeared' during his rule.

September 1998: The Government instigated a number of labour reforms, including the abolition of short-term contracts and the reduction of redundancy payments.

October 1998: Menem travelled to the United Kingdom for discussions with the Prime Minister, Anthony (Tony) Blair, regarding defence and trade, most notably the arms embargo imposed by the United Kingdom on Argentina.

February 1999: President Menem and his Chilean counterpart, Eduardo Frei Ruiz-Tagle, concluded a defence agreement.

March 1999: The Supreme Court rejected appeals that Menem be allowed to seek election to a third term in office.

July 1999: Following further talks between the British Government and the Falkland Islands' administration, the ban on the visit of Argentine citizens to the islands was revoked.

October 1999: A Spanish judge indicted 192 Argentine nationals on charges of genocide, terrorism and torture during the period 1976–83.

24 October 1999: Fernando de la Rúa Bruno of the UCR won 48.5% of the votes cast in the presidential election; the PJ candidate, Eduardo Duhalde, won 38.1% of the ballot. In concurrent elections to the Chamber of Deputies, an alliance of the UCR and Frepaso won 63 of the 130 available seats, the PJ 50 and the AR nine, leaving the Frepaso-UCR alliance with 127 of the 257 seats in the Chamber. De la Rúa was inaugurated in December, appointing Rodolfo Terragno as Chief of Cabinet.

December 1999: The new Congress approved a budget including a number of austerity measures, a tax-reform programme and a federal revenue-sharing scheme.

April 2000: Major revisions to the employment law enacted by the Senate provoked demonstrations by public-sector trade-unions; allegations that Senators had received bribes from government officials in order to approve the legislation were subsequently made.

September 2000: The Senate approved the repeal of legislation granting immunity from criminal investigation to legislators, judges and government ministers, thus allowing an investigation to be held into the allegations of bribes being paid to Senators.

October 2000: The Vice-President, Carlos Alvarez, resigned in protest at the failure of the President to remove, in the course of a government reorganization in which Terragno was replaced by Chrystian Colombo, two ministers who were accused of involvement in the corruption scandal.

March 2001: The Minister of the Economy, José Luis Machinea, resigned; he was replaced by Ricardo López Murphy, who effected major reductions in public expenditure.

7 June 2001: Menem was placed under house arrest, following charges of his involvement in illegal arms sales during his time in office; he was later prevented from leaving the country to go on honeymoon with his new wife.

19 September 2001: A federal court ordered the arrest of 18 Argentine citizens now resident in Spain, for their involvement in human-rights violations during the 'dirty war'.

27 September 2001: Videla was charged by a federal court of murdering political opponents during his dictatorship.

14 October 2001: In elections to Congress, the PJ won 37.4% of the votes cast, the UCR-Frepaso alliance 23.1%; the remaining votes were divided among a number of smaller parties.

Bahamas

c. AD **900:** Arawak Indians settled the islands after moving northwards from South America.

12 October 1492: The Italian navigator in the service of the Spanish Crown, Cristoforo Colombo (Christopher Columbus) discovered the Bahamas during his first voyage to the New World. Columbus named the largest islands San Salvador, Santa María de la Concepción, Fernandina and Isabella.

1500–13: Spanish explorers and traders hunted down and carried off an estimated 40,000 indigenous Arawaks to work as slaves in Cuba and Hispaniola.

1514: A Spanish explorer, Juan Ponce de León, visited the islands and found them to be uninhabited. He named the islands Baja Mar (low in the water), which became corrupted to their present name, the Bahamas.

1625: French settlers tried to establish a colony on the island of Great Abaco, but the attempt failed.

1629: England made its first formal claim to the islands.

1648: The nonconformist 'Company of Eleutherian Adventurers' landed in the Bahamas after leaving Bermuda, where they had encountered discrimination. The company, led by the former Governor of Bermuda, William Sayle, founded the first British colony in the islands.

1656: A second group of Puritan Christians arrived from Bermuda and settled on an island they named 'New Providence'.

1670: Control of the Bahamas was transferred from Bermuda to six Lord Proprietors of the Carolinas, who established large cotton plantations.

1691: Nicholas Trott of Bermuda was appointed Governor of the Bahamas. Trott built Fort Nassau (completed in 1695) to defend the harbour from pirate attacks.

1717: The British Crown assumed direct control of civilian and military affairs.

1718: The British King, George I, proclaimed the Bahamas a Crown Colony.

1729: A system of parliamentary government was created, with a representative House of Assembly.

May 1782: Spain briefly occupied the Bahamas.

1783: The Treaty of Paris returned the Bahamas to British sovereignty.

1834: The United Kingdom abolished slavery throughout its Empire. The Bahamas subsequently experienced a rapid decline in its economy and population, both of which were highly dependent on imported slaves.

1919–33: Rum-smugglers made extensive use of the islands during the time of Prohibition in the USA, causing an expansion of the economy and an increase in population to around 60,000.

1940–45: The Duke of Windsor served as Governor of the Bahamas.

August 1953: Lord Ranfurly replaced Maj.-Gen. Sir Robert Neville as the islands' Governor-General.

20 May 1960: Elections took place to four newly created seats in the House of Assembly. The opposition NPLP won all four, increasing its representation to nine of the 33 seats available.

23 February 1961: The House of Assembly approved a bill granting women the right to vote and to seek election to public office.

1–20 May 1963: A Constitutional Congress in the British capital, London, agreed that the House of Assembly would be expanded, while the Legislative Council would be renamed the Senate.

7 January 1964: The Bahamas was granted full internal self-government. The leader of the United Bahamian Party (UBP), Sir Roland Symonette, became the country's first Premier. The United Kingdom retained responsibility for foreign relations, defence and internal security.

10 January 1967: The Progressive Liberal Party (PLP) and the UBP each won 18 seats at elections to the 38-member House of Assembly. The PLP leader, Lynden (later Sir Lynden) Pindling, formed a Government after winning the support of an independent member.

1972: A general election was won by the PLP and Pindling was reappointed Prime Minister (as his office had been restyled in 1968).

10 July 1973: The Bahamas achieved full independence from the United Kingdom as the Commonwealth of the Bahamas Islands. Pindling became the country's first Prime Minister.

19 July 1977: The PLP obtained 55% of the votes cast at a general election; Pindling was reappointed Prime Minister.

10 June 1982: The PLP retained power after winning 32 seats in the newly enlarged 43-member House of Assembly. Pindling was reappointed Prime Minister, leading a virtually unchanged Cabinet.

8 October 1984: Two ministers resigned from the Cabinet, following their implication in financial irregularities and drugs-trafficking by a Royal Commission into government corruption. Pindling refused to resign, despite evidence that he had received several large gifts and loans from business executives.

19 June 1987: At legislative elections the PLP secured 31 of the 49 seats in the House of Assembly and Pindling was returned to power, despite opposition accusations of electoral malpractice.

19 August 1992: The opposition Free National Movement (FNM) won 33 of 49 seats at the general election, ending the PLP's record of five consecutive victories. The leader of the FNM, Hubert Ingraham, was appointed Prime Minister.

9 January 1995: The Deputy Prime Minister and Minister of Foreign Affairs, Orville Turnquest, was appointed Governor-General (the local representative of the nominal head of state, the British monarch), prompting a cabinet reshuffle.

14 March 1997: The FNM won a decisive victory at early legislative elections. obtaining 34 of the 40 seats in a reduced House of Assembly; the PLP secured the remaining six seats. Ingraham was reappointed Prime Minister. Following the announcement of the result, Pindling resigned as leader of the PLP; he was replaced by Perry Christie.

15 October 1998: Amid strong international criticism, two convicted murderers were executed. The executed men had had final appeals to the British Privy Council (which remained the ultimate court of appeal for a number of Caribbean states) rejected on the grounds that to keep them on 'death row' for more than five years in a jurisdiction where the death penalty was applicable was inhumane. The use of the death penalty had become a political issue in a number of Caribbean states where violent (often drugs-related) crime had increased in the 1990s, and remained highly popular, despite international advocacy of its abolition. Campaigners subsequently demanded the establishment of a court of similar standing, possibly having jurisdiction over a number of countries, in the Caribbean, to replace the Privy Council.

January 2000: The deputy leader of the PLP, Bernard Nottage, resigned his position and later established a rival party, the Coalition for Democratic Reform, thereby further weakening the PLP's position.

26–27 June 2000: The Organisation for Economic Co-operation and Development (OECD) and the Financial Action Task Force (FATF, part of the OECD) criticized the Bahamas' financial system, naming it as a tax haven. The OECD requested that the country reform its rules by 31 December 2005 or face sanctions.

26 August 2000: Sir Lynden Pindling died.

Barbados

c. 1600 BC: The first settlers to arrive on the island were the Arawaks, an Amerindian people who are believed to have arrived in longboats from what is now Venezuela.

c. AD 1200: The Caribs conquered the Arawaks and displaced them.

1536: Pedro a Campos, a Portuguese explorer, visited the island *en route* to Brazil. Campos named the island Os Barbados (the bearded ones) after the island's fig trees.

14 May 1625: A British expedition under Capt. John Powell claimed the island on behalf of King James I.

17 February 1627: Powell brought 80 settlers and 10 slaves to Barbados and officially colonized the island. The settlers named their settlement Jamestown, which later became known as Holetown, and produced tobacco, cotton and sugarcane using slave labour.

1639: King Charles I gave the Earl of Carlisle permission to colonize the island. The Earl's appointed Governor, Henry Hawley, founded a House of Burgesses (later House of Assembly), making the island the third parliamentary democracy in the world.

1640s: The highly successful sugar industry encouraged some 40,000 white farmers to settle in Barbados.

1650s: Following a decline in the importance of the sugar industry, many of the white settlers left the island.

1663: Barbados became a British Crown Possession.

1720: Barbados definitively lost its dominant role in the sugar industry with the emergence of the Leeward Islands and Jamaica as powerful commercial rivals.

1834: The United Kingdom abolished slavery throughout its Empire; Barbados subsequently introduced a four-year transitional period of 'semi-slavery'.

1876: A British proposal to create a confederation of Barbados and the Windward Islands led to violent unrest in Barbados. The proposals were subsequently abandoned.

1937: Protests at the poor economic climate and the high rate of unemployment escalated into violence. A British Royal Commission was established to investigate conditions.

April 1938: The Barbados Progressive League (BPL) was established, with Grantley (later Sir Grantley) Adams as its leader.

November 1946: At a general election the Barbados Labour Party (BLP, as the BPL had been renamed), led by Adams, won nine seats in the House of Assembly.

1 April 1947: A 10-year development plan worth £3.5m. was inaugurated. The plan was intended to improve the island's agriculture, education, public health, housing and town-planning, water supply, social welfare, industrial development and tourist industry.

February 1951: Full adult suffrage was introduced.

December 1951: At a general election the BLP obtained 16 of the 24 seats in the House of Assembly.

1 February 1954: The Governor of Barbados, Sir Robert Arundell, announced the introduction of a ministerial system of government under the premiership of Adams.

6 December 1956: Adams was returned to office at a general election, at which the BLP maintained its majority in the House of Assembly.

1958: The West Indies Federation was formed with Adams as Prime Minister, and Dr H. G. H. Cummins was appointed Premier of Barbados.

October 1961: Barbados achieved full internal self-government.

4 December 1961: At legislative elections, the Democratic Labour Party (DLP) won 14 of the 24 seats in the House of Assembly and the party's leader, Errol Barrow, became Premier.

27 February–5 March 1962: Representatives of Barbados and the seven Leeward and Windward Island colonies met in Barbados and requested that the British Government approve a new federation, known as the Little Eight (subsequently the Little Seven, following the withdrawal of Grenada), in the event of the dissolution of the West Indies Federation.

May 1962: The British Government formally dissolved the West Indies Federation.

30 November 1966: Barbados achieved full independence from the United Kingdom. Barrow became the island's first Prime Minister.

9 September 1971: At a general election the DLP secured 18 seats and the BLP six.

2 September 1976: The BLP defeated the DLP in a general election, and the BLP leader, John Michael Geoffrey (Tom) Adams, the son of Sir Grantley Adams, was appointed Prime Minister.

18 June 1981: At a general election the BLP, which had campaigned on its economic achievements, secured 17 of the 27 seats in the recently enlarged House of Assembly, and Adams was re-elected as Prime Minister.

11 March 1985: Adams died suddenly and was replaced as Prime Minister by his deputy, Bernard St John.

28 May 1986: The DLP won 24 seats in the House of Assembly, and the DLP leader, Barrow, became Prime Minister.

1 June 1987: Barrow died unexpectedly and was replaced as Prime Minister by his deputy, L. Erskine Sandiford.

1 February 1989: The former Minister of Finance, Richie Haynes, resigned from the DLP and founded the National Democratic Party (NDP), which became the main opposition party.

22 January 1991: At a general election the DLP retained 18 of the 28 seats in the enlarged House of Assembly, the BLP secured the remaining 10 seats. Sandiford remained Prime Minister.

7 June 1994: Sandiford lost a parliamentary vote of 'no confidence' following a number of controversial political appointments.

6 September 1994: Sandiford's DLP was comprehensively defeated by the BLP at legislative elections; the BLP secured 19 seats in the House of Assembly, compared with eight for the DLP. Owen Arthur, the leader of the BLP, was appointed Prime Minister.

December 1998: A report on the possible reform of Barbados' Constitution recommended the replacement of the British monarch (represented locally by the Governor-General) with a ceremonial President.

20 January 1999: The ruling BLP improved its majority substantially in elections to the House of Assembly, receiving 64.8% of the total votes cast and winning 26 of the 28 seats available. The DLP, despite securing 35.1% of votes, won only two seats. The BLP's electoral success was largely attributed to the party's successes in reviving the Barbadian economy. Arthur retained his position as Prime Minister and appointed a new Cabinet.

9 August 1999: The Minister of Health and the Environment, Elizabeth Thompson, was dismissed from the Cabinet.

15 April 2000: Arthur dismissed the Minister of Tourism and International Transport, George Payne, who was replaced by Noel Anderson Lynch.

26–27 June 2000: The Organisation for Economic Co-operation and Development (OECD) and the Financial Action Task Force (FATF, part of the OECD) criticized Barbados' financial system, naming it as a tax haven. The OECD requested that the country reform its rules by 31 December 2005 or face sanctions.

August 2000: Arthur announced that a referendum on the replacement of the monarchy with a republic would be held.

Belize

2500 BC–250 BC: The basic institutions of Maya civilization emerged in Central America, principally in the Yucatán Peninsula and what is now southern Mexico, Guatemala, Honduras and Belize.

AD 250–900: Cities flourished throughout what is now known as Belize, during the apogee of the Maya civilization.

1525: The Spanish conqueror of Mexico, Hernán Cortés, passed through the south-west of present-day Belize.

1530s: Montejo attempted to conquer the region for Spain, but encountered serious resistance from the Maya provinces of Chetumal and Dzuluinicob.

1600s: The area was occupied by the Spanish, and British buccaneers and woodcutters began to settle around Belize river.

1670: The Treaty of Madrid, signed by Spain and Great Britain, confirmed possession of countries and islands in the 'New World' already occupied by the latter. However, these colonies were not named and ownership of the coastal area between Yucatán and Nicaragua remained undefined.

1700s: Spanish forces repeatedly forced the British settlers (known as Baymen) to leave the area. However, the Spanish never settled in the region and the British always returned to expand their trade and settlement.

1763: At the conclusion of the Seven Years' War, the Treaty of Paris conceded to British settlers the right to cut and export logwood, but asserted Spanish sovereignty over the region. The area in which the woodcutters could operate remained legally undefined.

1779: Spanish forces captured the area and took Baymen and slaves to Yucatán. The slaves were freed after declaring loyalty to Spain. The Baymen were sent to Cuba.

1783: Under the terms of a further treaty, Spain recognized British rights to fell logwood in the region between the Hondo and Belize rivers.

1784: After British settlers returned to the region, the Governor of Jamaica, Col Edward Marcus Despard, was appointed as the first Superintendent of the settlement.

1786: The Convention of London expanded British logging rights southward to the Sibun river. The Convention also affirmed Spanish sovereignty over the region.

1798: Spain unsuccessfully attempted to remove British settlers from the area by force.

1802: Spain recognized British sovereignty over the region in the Treaty of Amiens.

1832: Large number of Garifuna (the descendants of Carib people of the Lesser Antilles and of Africans escaped from slavery), fleeing from civil war in Honduras, arrived in the territory.

1833: The act abolishing slavery throughout the British colonies was passed and slavery was progressively eliminated during the subsequent years; the practice is considered to have been discontinued in Belize (as the British territory had become known, after its major river) by 1838.

1847–53: Several thousand Spanish-speaking refugees settled in the north of the territory and Maya communities relocated to the north and west following the War of the Castes in Yucatán.

1859: The British-Guatemala Treaty was signed, defining the border with Belize. The British Honduras Company (BH Co, later Belize Estate Produce Company—BEC) was formed.

1862: The Settlement of Belize in the Bay of Honduras was formally declared a British Colony (it had been subject to British law since 1840) and named British Honduras. The Crown's representative was elevated to the rank of Lieutenant-Governor, subordinate to the Governor of Jamaica.

1871: British Honduras was declared a Crown Colony after the Legislative Assembly dissolved itself in 1870 and requested the establishment of direct British rule in return for greater security. A new Constitution was inaugurated and the Legislative Assembly was renamed the Legislative Council.

1893: Mexico renounced its long-standing claim to Belizean territory north of the Sibun river.

1933: Guatemala re-asserted its claim to the region on the basis that the conditions of the 1859 British-Guatemala Treaty had not been met by the United Kingdom.

1934: Antonio Soberanis Gómez led protests by the Labourers and Unemployed Association, demanding emergency relief and a minimum wage from the Government.

31 December 1949: In defiance of the wishes of the Legislative Council, the Governor significantly devalued the territory's currency, the British Honduras dollar.

1950: The anti-British, pro-independence People's United Party (PUP) was established under the leadership of George Price, as a direct consequence of the currency devaluation.

28 April 1954: Following the establishment of a new Constitution granting universal adult suffrage, a general election was won by the PUP, which secured 66.3% of the vote and obtained eight of the nine elected seats in the new Legislative Assembly.

21 March 1957: The PUP won all nine seats at a general election.

1 March 1961: Under a ministerial system of government established by a new Constitution, Price was appointed First Minister after the PUP won all 18 seats in an enlarged legislature, restyled the House of Representatives.

1 January 1964: British Honduras was granted internal self-government; the United Kingdom retained responsibility for defence, external affairs and internal security.

1 March 1965: Following an election in which the PUP claimed 16 of the 18 seats in the House of Representatives, Price was appointed Premier of British Honduras.

May 1968: All parties in British Honduras rejected the 'Webster Proposals'—a draft treaty presented by a US Commission mediating the dispute between Guatemala and the United Kingdom—which granted Guatemala as much control over the territory as was exercised by the United Kingdom.

1969: At a general election the PUP extended its representation in the House of Representatives to 17 seats.

1971: British Honduras joined the Caribbean Free Trade Area (CARIFTA).

1 January 1972: The new town of Belmopan replaced Belize City as the capital.

1 June 1973: British Honduras was officially renamed Belize.

30 October 1974: A general election resulted in victory for the ruling PUP, but with a reduced majority. PUP candidates won 12 out of 18 seats in the House of Representatives and Price remained Premier.

21 November 1979: The PUP was again returned to office at a general election. The party, led by Price, obtained 51.8% of the votes cast and won 13 of the 18 seats in the legislature.

March 1981: Guatemala accepted Belizean independence in exchange for access to the Caribbean Sea through Belize.

21 September 1981: Belize attained full independence with George Price as Prime Minister and a new Constitution came into effect. However, Guatemala refused to recognize the country. The United Kingdom agreed to continue its defence commitments for a period of time, principally to defend the country against Guatemalan territorial claims. Moreover, also in that month, Belize was admitted to the Commonwealth and the UN, and applied for full membership of the Non-Aligned Movement (Belize had enjoyed 'special status' within the movement since 1976).

January 1983: Belize rejected Guatemala's proposal that Belize should cede the southern part of the country.

14 December 1984: Manuel Esquivel of the centre-right United Democratic Party (UDP) became Prime Minister after defeating Price's PUP at a general election, thus ending the PUP's uninterrupted 30-year period in government. The UDP won 21 of the 28 seats in the House of Representatives.

4 September 1989: Price and the PUP returned to power following general elections in which the PUP won some 51% of votes cast and 16 of the 28 seats in the House of Representatives.

5 September 1991: Guatemala recognized Belize as a sovereign and independent state. Diplomatic relations with Guatemala were re-established, although the country's claim over approximately one-half of Belizean territory remained. As a consequence, the United Kingdom announced its intention to withdraw its troops by 1994.

April 1993: Belize and Guatemala signed a non-aggression pact.

13 May 1993: It was announced that almost all British forces were to be withdrawn from Belize.

30 June 1993: Manuel Esquivel became Prime Minister after an alliance of the UDP and the National Alliance for Belizean Rights (NABR) won 16 seats in the House of Representatives (which had been enlarged by one seat), compared with 13 secured by the PUP.

20 July 1993: Esquivel suspended the agreement reached with Guatemala in September 1991, on the grounds that it made too many concessions in return for Guatemala's recognition of Belizean independence.

17 November 1993: Sir Colville Young was appointed as Governor-General (the local representative of the nominal head of state, the British monarch).

1 March 1994: Guatemala formally reaffirmed its territorial claim to Belize.

November 1997: The PUP presented proposals for the reform of the political system, including the establishment of a republican form of government, with the Governor-General to be replaced by a president elected by the National Assembly. However, the proposals were rejected by the Government.

27 August 1998: The PUP secured an overwhelming victory at a general election, winning 59.4% of votes cast and 26 parliamentary seats; the UDP obtained the remaining three seats. The leader of the PUP, Said Musa, was appointed Prime Minister.

January 1999: A Political Reform Commission was established to review the system of governance in Belize.

July 1999: Musa rejected allegations that the treasurer of the British Conservative Party, Michael Ashcroft, who had extensive business and banking interests in Belize and was its Ambassador to the UN, had used improper influence in Belizean affairs and that he might have been involved in 'money-laundering'. Ashcroft resigned his ambassadorial position in March 2000.

January 2000: Belizean security personnel killed a Guatemalan civilian, claiming that they had acted in 'self-defence'.

February 2000: Guatemala announced that it would pursue its territorial claim to an area of Belize through international courts. In the same month a bomb exploded at the Guatemalan embassy in Belmopan.

8 March 2000: The Belizean ambassador was expelled from Guatemala after it was revealed that he had assisted the flight from the country of three Belizean border guards who had been detained in Guatemala. The development significantly increased the tensions between the two countries.

26–27 June 2000: Belize was included on a list of harmful tax havens, published by the Organisation for Economic Co-operation and Development (OECD) and the Financial Action Task Force (FATF, part of the OECD).

17 January 2001: The Belizean and Guatemalan Governments adopted a new plan to resolve the border tension between the two countries. The proposal urged clarification of the procedure for removing Guatemalan settlers in western Belize and for an independent body to establish the exact co-ordinates of the disputed border.

8 October 2001: Towns and infrastructure in the south of the country were severely damaged by Hurricane Iris. The north of the country had been similarly damaged by Hurricane Keith in October 2000.

Bolivia

***c.* 600 BC:** The Tiahuanacan Empire, the first great Andean empire, began to extend its influence throughout much of modern-day Bolivia from the city of Tiahuanaco near Lake Titicaca.

***c.* AD 1200:** The Tiahuanacan Empire collapsed suddenly and was replaced by seven regional kingdoms, the inhabitants of which belonged to the Aymará ethnic group.

1400s: The southern Aymará kingdoms began to decline in power and were conquered by the Inca Empire, the capital of which was Cuzco (now in Peru). The Incas made the Bolivian highlands one of the four administrative units of the Empire and ruled through the traditional Aymará nobility.

1470: Several Aymará kingdoms rebelled against Inca rule, but were defeated. The Incas consolidated their control by granting land in the area to loyal Quechua-speaking communities from the Peruvian highlands.

1532: A Spanish expedition led by Francisco Pizarro captured and executed Inca Atahualpa, effectively ending the Inca Empire and beginning Spanish domination of the central Andes.

1538: Pizarro defeated the Inca forces in the Titicaca region.

1548: The Spanish Crown re-established control of the Bolivian highlands (known as Upper Peru) following the end of a civil war between supporters of Pizarro and of his lieutenant, Diego de Almagro. The colonial authorities founded the city of La Paz as a commercial centre and began to mine silver deposits at Potosí.

1570s: The Spanish Viceroy, Francisco de Toledo, introduced forced labour in the silver mines. Few of the Quechua or Aymará population survived their term underground.

17th century: Silver production decreased dramatically following the exhaustion of the most accessible deposits. Moreover, the system of forced labour had significantly reduced the indigenous population.

1650s: Forced labour and technological improvements significantly increased production at Potosí, which at the time had the largest population of any city in the Western Hemisphere.

1780s: Revolts against Spanish rule were organized by the indigenous leader, Túpac Amaru II, principally in Upper Peru.

1809: The authorities defeated an uprising by Upper Peru's political elite (the *criollos*), who refused to recognize the new Spanish authorities and demanded independence for the territory. Later that year Pedro Domingo Murillo led a further *criollo* revolt and proclaimed Upper Peru an independent state. Forces sent from Peru and Argentina suppressed the rebellion, but failed to re-establish Spanish control over the entire territory.

1 April 1825: The royalist leader in Upper Peru, Gen. Pedro Antonio de Olañeta, was defeated by rebel forces under the leadership of Antonio José de Sucre, the general of the Venezuelan 'Liberator', Simón Bolívar.

6 August 1825: A proposed union of Upper Peru with Argentina or Peru was rejected and a declaration of independence issued, naming the country the Republic of Bolivia, in honour of Bolívar. Bolívar was appointed the country's first President, and he rapidly abolished the system of forced labour and granted land to the indigenous people.

January 1826: Bolívar resigned as President and was replaced by Sucre.

May 1829: One of Bolívar's former lieutenants, Andrés de Santa Cruz y Calahumana, was sworn in as Bolivia's first native-born President. Santa Cruz introduced the first civil and commercial codes in Latin America and created economic and political stability.

1836: The Peru-Bolivia Confederation was established with the intention of resisting attempts at expansion by Chile and Argentina, both of which subsequently declared war on the Confederation.

20 January 1839: Chilean forces defeated the Confederation army, precipitating the dissolution of the Confederation.

1841–47: The presidency of Gen. José Ballivián y Segurola restored political stability to the country and encouraged free trade.

1847: Ballivián y Segurola was overthrown and Manuel Isidoro Belzú Humérez became President.

1855: After surviving at least 42 attempted *coups d'état*, Belzú finally resigned after losing the support of the political élite. His son-in-law, Gen. Jorge Córdova, replaced him as President.

1857: José María Linares Lizarazu overthrew Córdova and became Bolivia's first civilian President. Increased foreign investment and increased efficiency in the mining sector contributed to steady economic growth during his presidency.

1861: Linares was ousted by a military *coup d'état* led by Gen. José María Achá Valiente, who executed Córdova and 71 other people considered loyal to Belzú.

1864: Gen. Mariano Melgarejo Valencia overthrew Gen. Achá. His regime subsequently became notorious for its corruption, economic mismanagement and ill-treatment of the indigenous population.

1867: Gen. Melgarejo signed a treaty with Brazil in which he ceded 102,400 sq km of Bolivian territory in exchange for improved access rights to the Atlantic Ocean.

1871: Agustín Morales Hernández overthrew Melgarejo, but retained his predecessor's autocratic style of government.

1873–76: Following the assassination of Morales by his nephew, a number of short-lived civilian and military leaders failed to establish themselves in the Presidency.

1876: A high-ranking military officer, Hilarión Daza Groselle, seized power and established a repressive regime that attracted widespread national opposition.

14 February 1879: Chile invaded Bolivia after the latter failed to observe the terms of a treaty between the two countries establishing their respective boundaries. The following month Bolivia, in alliance with Peru, declared war on Chile; however, the Bolivian forces were defeated and Daza was obliged to resign. Gen. Narciso Campero Leyes replaced him as President.

1880: The War of the Pacific ended when Chile decisively defeated the Bolivian and Peruvian joint forces. Bolivia withdrew all its forces from its coastal territory, which came under Chilean control.

1884: Campero held presidential elections, that were won by a mining magnate, Gregorio Pacheco Leyes.

1888: The conservative candidate, Aniceto Arce Ruíz, was elected President.

1899: The Partido Liberal (PL) overthrew the Government following a dispute over regionalism and federalism, and the leader of the party, José Manuel Pando, became President.

1900–03: A secessionist uprising in the eastern province of Acre obliged Bolivia to cede control of the area to Brazil. Under the terms of the Treaty of Petropolis, Bolivia relinquished 191,000 sq km of territory in exchange for 5,200 sq km of Brazilian territory and compensation of US $10m.

1904: The PL candidate, Ismael Montes Gamboa, was elected President. Bolivia and Chile signed the Treaty of Peace and Friendship by which Bolivia officially ceded its coastal provinces to Chile, thereby losing its access to the Pacific.

1914: The Partido Republicano (PR) was formed to campaign against political abuses and the loss of national territory.

1917: The Liberal candidate, José Gutiérrez Guerra, defeated a Republican candidate in a presidential election.

1920: The PR seized power in a peaceful *coup d'état* and installed Bautista Saavedra Mallea as President.

1926: Hernando Siles Reyes became President.

1930: Siles was overthrown by the military after seeking election to a second consecutive term as President, in contravention of the Constitution.

March 1931: The ruling military junta held presidential elections, that were won by Daniel Salamanca Urey.

1932–34: A series of incidents on Bolivia's border with Paraguay escalated into the Chaco War. Bolivia lost every major battle and President Salamanca was obliged by the armed forces to resign after he announced his intention to continue with the war. The former Vice-President, José Luis Tejada Sorzano, replaced him as President.

14 June 1935: A multi-national commission brokered an armistice between Bolivia and Paraguay. Under the terms of the final peace agreement, Bolivia was obliged to cede all of its territory in the Chaco region to Paraguay.

17 May 1936: Col David Toro Ruilova overthrew Tejada in a military *coup d'état*. Toro's programme of 'military socialism' was intended to extend social and economic justice in Bolivia, partly by nationalizing foreign-owned companies.

1937: Col Germán Busch Becerra overthrew Toro with the support of a group of radical army officers.

1938: Busch introduced a new Constitution, which emphasized the primacy of the common good over private property and favoured government intervention in social and economic relations. It also legally recognized the Indian communities.

1939: Frustrated by his inability to bring about change, Busch committed suicide. Busch's Chief of Staff, Gen. Carlos Quintanilla Quiroga, was appointed acting President.

1940: Gen. Enrique Peñaranda Castillo was elected President.

December 1943: An alliance of military officers and the far-right Movimiento Nacionalista Revolucionario (MNR) overthrew Peñaranda, and Maj. Gualberto Villarroel López became President. Villarroel appointed several senior MNR figures, including the party's leader, Víctor Paz Estenssoro, to his cabinet, however they later resigned after the USA refused to recognize the Government.

1944: The MNR won a majority at legislative elections; the USA subsequently agreed to recognize a Government including MNR ministers.

21 July 1946: The Villaroel Government was overthrown by an uprising led by the conservative 'Popular Revolutionary Movement' and Paz Estenssoro went into exile. Over 1,000 people were killed in the revolution. Nestor Guillén was named President of the Revolutionary Junta.

5 January 1947: Dr Enrique Hertzog, a former opponent of Villaroel, was elected President and formed a socialist Government.

22 October 1949: Dr Hertzog retired on grounds of ill health and was replaced by Mamerto Urriolagoitia, hitherto the Vice-President.

11 April 1950: The Partido Comunista was banned following allegations by the Government that party activists were organizing a revolution.

May 1951: The MNR candidate, Paz Estenssoro, who returned from exile in Argentina to contest the presidential election, received the greatest number of votes but failed to secure an absolute majority. President Urriolagoitia refused to acknowledge the legitimacy of the election results and granted power to a military junta before going into exile. The junta installed Gen. Hugo Ballivián as President and declared a national state of siege.

6–9 April 1952: The military junta was overthrown by a popular uprising led by the MNR, during which several hundred people were killed.

15 April 1952: Paz Estenssoro returned from exile and assumed the presidency.

17 June 1956: The MNR candidate, Hernán Siles, won a presidential election, after Paz Estenssoro decided not to seek re-election. The MNR also performed strongly at legislative elections.

19 April 1959: The far-left Falange Socialista Boliviana (FSB) unsuccessfully attempted to overthrow the Government.

5 June 1960: At a presidential election, Paz Estenssoro was elected to office with a large majority.

31 May 1964: Paz Estenssoro won 70% of the votes cast in a presidential election boycotted by all the opposition candidates, who had expressed discontent at the timing of the elections.

20 September 1964: Martial law was declared by the ruling authorities and former President Siles and a number of other prominent politicians were arrested after a plot to overthrow the Government was alleged.

4 November 1964: The Government was overthrown by a military *coup d'état* led by the former Vice-President, Gen. René Barrientos Ortuño, who proclaimed himself President and formed an entirely military Government. Paz Estenssoro went into exile.

26 May 1965: Gen. Alfredo Ovando Candía, the Commander-in-Chief of the armed forces, became Co-President of the ruling military junta.

4 January 1966: Gen. Barrientos resigned following a series of demonstrations by tin-miners, leaving Gen. Ovando in power.

3 July 1966: Gen. Barrientos was elected President after standing as the candidate of the armed forces and of the Frente de la Revolución Boliviana (FRB).

27 March 1967: The Bolivian Government announced that 'foreign forces' had entered Bolivia and were preparing a revolution. The group, led by the Argentine revolutionary leader, Ernesto 'Che' Guevara, subsequently inflicted serious losses on the armed forces in the area.

8 October 1967: The Bolivian army captured and executed Guevara, whose guerrillas had failed to attract the support of the local population.

25 July 1968: The Cabinet resigned *en masse* in protest at a number of political arrests carried out by the security forces, and Barrientos appointed a military Cabinet.

27 April 1969: President Barrientos was killed in a helicopter accident, and was succeeded by the then Vice-President, Luis Adolfo Siles Salinas.

26 September 1969: President Siles was overthrown by a military *coup d'état* led by Gen. Ovando, who subsequently established a Revolutionary Government.

6 October 1970: Gen. Ovando resigned after losing the support of both right- and left-wing military officers. Gen. Juan José Torres was sworn in as his replacement.

22 August 1971: Col (later Gen.) Hugo Bánzer Suárez led a successful *coup d'état* against Gen. Torres, with the support of the FSB, the army, and a section of the MNR. At least 120 people were killed in the right-wing uprising.

7 November 1974: An attempted military *coup d'état* in Santa Cruz was defeated by Bánzer, and two days later Bánzer announced he was indefinitely postponing presidential and legislative elections and that the military would remain in power until 1980.

2 August 1977: The indigenous Quechua and Aymará languages were designated official languages for the first time since the Spanish conquest.

9 July 1978: Presidential and legislative elections were held, but were subsequently annulled following allegations of fraud.

20–21 July 1978: The armed forces' candidate for the presidency, Gen. Juan Pereda Asbún, led a military uprising, obliging Bánzer to concede defeat and resign his position.

24 November 1978: President Pereda was overthrown by a military *coup d'état* led by the Commander-in-Chief of the Army, Gen. David Padilla Arancibia. Gen. Padilla pledged to hold early elections and transfer power to an elected Government.

1 July 1979: No single candidate won a majority at the presidential election, contested by two former holders of the office, Siles Zuazo and Paz Estenssoro, representing rival factions of the MNR. Centre-left and moderate rightist parties won the majority of seats at legislative elections.

6 August 1979: In the absence of a clear victor, Congress elected Walter Guevara Arce as interim President, pending a new presidential election.

1 November 1979: Col Alberto Natusch Busch led a military *coup d'état*, overthrowing Guevara's interim presidency.

16 November 1979: Faced by overwhelming hostility to his regime, Col Natusch resigned as President. Congress appointed the President of the Chamber of Deputies, Lidia Gueiler Tejada, as the interim President.

29 June 1980: At a presidential election the centre-left candidate, Siles Zuazo, received the greatest number of votes, but failed to win an outright majority.

17 July 1980: A military junta, led by Gen. Luis García Meza, overthrew the interim Government and assumed power.

4 August 1981: Gen. García Meza resigned control to a military junta after army rebels took control of Santa Cruz.

4 September 1981: A member of the junta, Gen. Celso Torrelio Villa, was appointed President.

19 July 1982: President Torrelio resigned amid rumours of an impending *coup d'état*. He was replaced as President by the Chief of Staff of the Army, Gen. Guido Vildoso Calderón.

10 October 1982: President Calderón resigned, amid a severe economic crisis; the military junta announced that it would recognize the results of the disputed 1980 elections and appoint Siles Zuazo as President.

21 April 1983: García Meza was sentenced *in absentia* to 30 years' imprisonment after being found guilty of murder, human-rights abuses, corruption and fraud.

30 June 1984: Members of the armed forces, led by Col Rolando Saravia Ortuño, abducted President Siles Zuazo, but eventually surrendered.

14 July 1985: At a presidential election held one year earlier than scheduled, Bánzer obtained the greatest share of the votes cast, with Paz Estenssoro in second place. Paz Estenssoro was successful in the vote in Congress between the two most popular candidates to decide the presidency.

22 January 1986: The President effected a major cabinet reshuffle intended to consolidate the Government's economic reforms.

14 July–15 November 1986: With the approval of the Bolivian Government, US troops carried out 'Operation Blast Furnace', which sought to eradicate the majority of northern Bolivia's coca crop, considered a major source of the cocaine available in the USA.

6 December 1987: Municipal elections were held to 2,400 local-government seats. The leftist Movimiento de la Izquierda Revolucionaria (MIR) and the right-wing Acción Democrática Nacionalista (ADN) significantly outperformed the President's MNR.

7 May 1989: At presidential elections, none of the three main presidential candidates—Bánzer representing the ADN, Jaime Paz Zamora of the MIR and Gonzalo Sánchez de Lozada of the MNR—obtained an outright majority.

5–6 August 1989: Congress elected Paz Zamora as President, after he agreed to the creation of a political alliance with Bánzer and the ADN.

8 August 1991: President Paz Zamora effected a major cabinet reshuffle, which was reportedly intended to reduce corruption at ministerial level.

24 January 1992: Bolivia and Peru signed an agreement designating the Peruvian port of Ilo as a free zone intended primarily for Bolivian use.

18 March 1992: President Paz Zamora carried out a major cabinet reshuffle intended to combat his Government's significant decline in popularity.

13 May 1992: The country's largest labour organization, the Central Obrera Boliviana (COB), called for a popular uprising to overthrow the Government, which had become increasingly unpopular as a consequence of its privatization and coca-eradication programmes.

25 March 1993: The Government reached an agreement on higher wages with the COB following several weeks of violent unrest by workers and students.

6 June 1993: The MNR candidate, Sánchez de Lozada, won the largest share of the votes cast at a presidential election, and was elected President following the withdrawal of Bánzer, the second-placed candidate.

25–26 February 1994: Around 15,000 coca growers demonstrated against the Government's coca-eradication programme. The authorities subsequently agreed to focus on destroying coca refineries rather than plantations.

26 March 1994: Paz Zamora resigned from public life following the publication of a report suggesting that he had knowingly suppressed evidence of drugs-trafficking.

10 January 1995: Paz Zamora declared himself innocent of all the charges against him and announced his return to public life.

18 April 1995: The Government arrested some 370 trade-union leaders and declared a state of siege following several weeks of industrial action and street demonstrations against proposed educational reforms and the privatization of state-owned companies.

18 July 1995: The Government extended the duration of the state of siege by a further 90 days in response to continued labour unrest and protests by coca growers.

February 1996: Ten MNR members of Congress resigned and 12 others were suspended on criminal charges following reports of abuse of the personal expenses system.

1 June 1997: The ADN candidate, Bánzer, won the greatest share of the votes cast (22.3%) at the presidential election. Juan Carlos Durán, representing the MNR, secured 17.7% and Paz Zamora (MIR) 16.7%. At concurrent legislative elections the ADN secured 33 of the 130 seats in the Chamber of Deputies (compared with 26 won by the MNR, 25 by the MIR, 21 by the populist Unión Cívica Solidaridad—UCS—and 17 by the Conciencia de Patria—Condepa) and 13 of the 27 in the Senate (the MIR won six, the MNR and Condepa three each and the UCS two). The ADN subsequently concluded a pact with the MIR, the UCS and Condepa to form a congressional majority.

5 August 1997: Bánzer was elected President by the National Congress.

April 1998: Coca producers clashed with the security forces on a number of occasions following the implementation by the Government of an ambitious coca-eradication plan, known as the 'Dignity Plan', which aimed to eradicate all illegal coca plantations by 2002.

3 August 1998: President Bánzer expelled Condepa from the ruling coalition following the emergence of serious divisions within the party leadership. A major cabinet reshuffle was subsequently announced.

2 September 1998: A large number of coca growers held a march in La Paz to protest at the Government's eradication plans.

31 March 1999: The Minister of Labour, Leopoldo López, resigned after he was implicated in a contraband scandal.

31 May 1999: A new penal code came into effect, which incorporated the customary law of the Quechua and Aymará populations into Bolivian law for the first time.

21 June 1999: President Bánzer carried out a major cabinet reshuffle following the resignation of the Minister of the Interior, Guido Nayor.

3 February 2000: President Bánzer dismissed the Nueva Fuerza Republicana (NFR) from the ruling coalition after the party's leader expressed support for recent violent demonstrations in Cochabamba against sharp increases in water-rates.

8–20 April 2000: The Government declared a state of emergency following violent clashes between the security forces and peasant farmers. Demonstrations against increased water-rates persisted in Cochabamba.

September 2000: The country was effectively brought to a standstill by road-blocks and demonstrations when teachers, striking to demand higher salaries, were joined by peasants protesting against plans by the Government to tax water used for irrigating crops. Moreover, coca growers continued with their protests against government plans to restrict the sale of the coca leaf. At least 10 people died in clashes between police and protesters.

6 October 2000: Demonstrators and government representatives reached an agreement to end the strikes and disturbances. The Government pledged to increase salaries for teachers and public-sector workers and to invest in projects to assist farmers to diversify away from coca production.

20 October 2000: President Bánzer reshuffled the Cabinet and created a new Ministry of Peasants', Indigenous Peoples' and Ethnic Affairs.

14 January 2001: The Minister of the Presidency, Walter Guiteras, resigned following allegations that he had attempted to bribe police officers.

April 2001: A number of demonstrations and strikes took place in protest at the Government's alleged failure to implement the terms of the agreement signed in October 2000.

21 May 2001: The Government and protesters' representatives agreed to establish seven bilateral commissions to review various aspects of government policy.

7 August 2001: Bánzer resigned the presidency on grounds of ill health; the Vice-President, Jorge Quiroga, succeeded him. Quiroga pledged to solve the country's economic crisis and to reduce poverty and corruption.

Brazil

c. **1000 BC:** The indigenous Tupí-Guaraní people inhabited an area extending from the east coast of South America to the Andes; they were nomadic and lived by hunting and gathering.

AD 1494: Portugal and Spain concluded the Treaty of Tordesillas, by which the known world was divided along a line of longitude established 370 leagues west of the Cape Verde Islands. Spain could claim the territory which lay west of the line and Portugal that which lay to the east.

22 April 1500: A Portuguese fleet, led by the navigator Pedro Alvares Cabral, arrived on the east coast, which Cabral named Terra da Vera Cruz (Land of the True Cross). Portuguese traders were subsequently attracted to the land by the abundance of brazilwood trees (which produced a valuable red dye) and established trading posts.

1530: A second Portuguese expedition, led by Martin Afonso de Sousa, founded the settlement of São Vicente (near the present-day city of Santos).

1536: The King of Portugal divided his newly acquired territory into a number of captaincies, which he granted to noblemen—known as *donatarios*—to develop and defend. The donatarios introduced sugar-cane cultivation and cattle to the region.

1549: A further fleet, under the command of Tomé de Sousa, brought colonists, soldiers, priests, royal officials and Roman Catholic Jesuit missionaries. A capital for the colony was established at Salvador.

1550s: Attempts by the sugar-plantation owners to recruit the indigenous peoples as labour failed owing to native resistance and the deaths of thousands from smallpox. Slaves were subsequently imported from Africa.

1570s: The Jesuits attempted to convert the indigenous peoples to Christianity and organized them into villages, called *aldeias*.

1620s: The Dutch attacked the settlements of Rio de Janeiro, Salvador and Recife.

1630s–50s: Recife and Olinda were occupied by Dutch forces. At this time, competition from the expansion of sugar plantations on the Caribbean islands also caused the sugar trade in the region to decline.

late 17th century: Gold was discovered by Portuguese adventurers exploring the mountain streams in the area north of the Rio de Janiero. In the first half of the next century, several thousand prospectors migrated to the area. The King of Portugal decreed that the crown would receive one-fifth of the gold minted in Brazil.

1720s: Deposits of diamonds were found in the region north of the gold-mining area; together with gold and sugar, the precious stone became one of the colony's main export commodities.

1750: Spain and Portugal concluded the Treaty of Madrid, which moved the 'Tordesillas Line' to the west to constitute the country's present-day western borders.

1755: The chief minister of the King of Portugal, the Marquis de Pombal, introduced legislation prohibiting the enslavement of the indigenous tribes and declaring them to be free citizens of the colony. The Marquis also sought to appoint Brazilian-born colonists to key administrative posts and enacted a number of economic reforms. The opposition of the Jesuits to these reforms resulted in their expulsion from the colony.

1763: The colonial authorities transferred the territory's capital from Salvador to Rio de Janeiro.

1789: A group of noblemen in the captaincy of Minas Gerais led a rebellion against what they perceived as increased imperial control and the imposition of new taxes. The revolt was suppressed and the dissidents imprisoned or exiled; Joaquim José da Sliva Xavier (Tiradentes), the only non-noble rebel, was hanged.

1808–21: Following the invasion of the Portuguese capital, Lisbon, by troops of the French Emperor, Napoléon Bonaparte, the Portuguese royal family fled to Rio de Janeiro. From there, the regent, Dom João VI, administered the Portuguese Empire, which included territories in Asia and Africa, in addition to the American possessions.

1815: The colony was promoted to the status of kingdom.

1817: A republican rebellion occurred in Pernambuco, which soon provoked unrest in other states; Dom João banned all secret societies and ordered troop reinforcements to be sent to Rio de Janiero, Salvador and Recife.

1820: High-ranking members of the Portuguese army rebelled against the rule of Dom João and demanded a constitutional government. They offered the regent a role as the constitutional monarch of the Empire, on condition that he returned to Brazil.

1820s: Coffee began to be cultivated in the states of Minas Gerais, Rio de Janeiro and São Paulo.

1821: Following the capture of the provincial governments of Bahia and Belém by officers and rebel troops, soldiers in Rio de Janeiro forced Dom João to take an oath to abide by any constitution that the Parliament drafted. The regent fled to Portugal, designating his son, Pedro, as his successor.

7 September 1822: Following attempts by the Portuguese Parliament to reduce the powers of Pedro in Brazil, the regent declared the colony's independence and installed himself as King Pedro I. He subsequently employed the services of a former British naval officer, Lord Thomas Cochrane, to suppress separatist unrest.

1824: Pedro I closed the Constituent Assembly after it sought to limit his powers in a new constitution. The king oversaw the drafting of a constitution, in which the ultimate legislative authority lay with the monarch.

1825–28: A dispute with Argentina over the southern border and sovereignty over Uruguay led to the Cisplatine War; the cost of the conflict caused unrest in the kingdom.

August 1825: Portugal recognized the independence of Brazil.

1831: Pedro I was forced to abdicate by officials suspicious of the number of Portuguese-born ministers in his cabinet. His four-year-old son was designated as his successor and a Council of Regents was appointed by the Parliament to govern on a temporary basis.

1835–45: A republican rebellion, known as the War of the Farrapos, took place in the states of Santa Catarina and Rio Grande do Sul.

1840: Owing to conflict between liberal and conservative factions in parliament, Pedro II was crowned at thirteen years of age.

1850: Under pressure from the United Kingdom, the Government was forced to prohibit the import of slave labour from Africa.

1864–70: The kingdom united with Argentina and Uruguay in a conflict against Paraguay, known as the War of the Triple Alliance. The three allies suffered a humiliating defeat.

1871: The children of slaves were emancipated.

13 May 1888: Princess Isabel, who was acting as regent for her aged father, approved the 'Golden Law', which abolished the institution of slavery. The plantation owners subsequently recruited increasing numbers of immigrants from Europe.

15 November 1889: High-ranking army officers led a *coup d'état* and forced the royal family to flee to Portugal. Gen. Manuel Deodoro da Fonseca was elected as the President of a provisional Government.

1890: Following a number of disputes with Parliament, Gen. Deodoro dissolved the Government and ruled by decree.

1891: Deodoro resigned; he was replaced by Floriano Peixoto. A new Constitution was adopted, establishing a federal system for the country, thenceforth known as the United State of Brazil, and providing for an elected Congress, comprising a House of Representatives and a Senate.

1894: Prudente de Morais was elected the first President of the federal republic.

1915–17: A rebellion took place in Constesdo against the despotism of the landowning élite; the revolt was inspired by a local healer, Miguel Lucena Boaventura.

1917: General strikes occurred in Rio de Janeiro, São Paulo, Recife, Santos and Porto Alegre.

July 1922: A group of junior military officers led an unsuccessful revolt in Rio de Janeiro.

1924: A further rebellion was attempted by members of the armed forces in São Paulo.

1929: During the global economic depression precipitated by the collapse of the stock markets in New York (USA), the coffee industry suffered from a fall in demand, causing hardship in Brazil, where coffee was the dominant export crop.

March 1930: Presidential elections were won by the government candidate, Júlio Prestes, who gained most of his support among the electorate of the larger states. His opponent was Getúlio Vargas, who represented a coalition of opposition parties and urban and military dissidents.

November 1930: Following a revolt among supporters of Vargas, President Prestes was forced to resign and Vargas was appointed the leader of a provisional Government.

1934: The Constituent Assembly elected Vargas to the presidency. A new Constitution enacted a number of reforms, including the introduction of suffrage for women.

1935: A revolt by the Communist movement was suppressed.

1937: Although he was re-elected President, Vargas was assisted by the armed forces in abrogating the Constitution. The Constituent Assembly was dissolved and Vargas proclaimed the establishment of the Estado Novo, which was to be organized on corporatist principles. The President proceeded to impose protectionist policies and also enacted social-welfare reforms.

1939–45: President Vargas took the country into the Second World War on the side of the Allies. Troops were sent to assist the Allied campaign in Italy; the USA also provided the Government with financial aid to facilitate the expansion of industry, especially the manufacture of war supplies.

October 1945: Several army units forced Vargas to resign following a *coup d'état*. In presidential elections, the candidate of the Partido Social Democrático (PSD), Eurio Dutra, received 55% of the votes cast.

January 1946: A new Constitution was promulgated, providing for the decentralization of power and regular elections.

1950: With the support of the Partido Trabalhista Brasileiro (PTB), Vargas was re-elected President; he subsequently authorized the establishment of federal-financed banks and corporations and introduced further social-welfare policies favourable to the working classes.

24 August 1954: Following a dispute with opposition parties and the military, intensified by allegations that his bodyguard had attempted to assassinate one of his opponents, Vargas committed suicide.

1955: A coalition of the PSD and the PTB elected Juscelino Kubitschek as President. During his time in office, Kubitschek supervised the expansion of heavy industry and the development of a basic infrastructure.

April 1960: The country's new federal capital, Brasília, was inaugurated.

October 1960: Jânio Quadros, representing the União Democrática Nacional (UDN), was elected President. João Goulart of the PTB was elected Vice-President.

July 1961: Quadros resigned, alleging a lack of support within Congress.

September 1961: Goulart acceded to the office of President; the Constitution was subsequently amended to provide for the creation of an office of Prime Minister.

January 1963: Following a succession of three Prime Ministers in 16 months, the abolition of the office was approved in a national referendum.

April 1964: The Chief of Staff of the armed forces, Gen. Umberto Castelo Branco, led a *coup d'état*; he was subsequently elected President by Congress.

October 1965: President Branco assumed full personal powers and abolished all the political parties.

December 1965: Two political parties were 'artificially' created, the pro-Government Aliança Renovadora Nacional (ARENA) and the opposition Movimento Democrático Brasileiro (MDB).

October 1966: Marshal Artur da Costa e Silva, who had been designated by Branco as his successor, was elected President, representing ARENA.

March 1967: A new Constitution was promulgated, renaming the country the Federative Republic of Brazil.

September 1969: Owing to ill health, President da Costa e Silva resigned; he appointed a triumvirate of military leaders to rule in his place.

October 1969: The Constitution was amended to invest executive authority in an indirectly elected President. The Congress was reinstated and Gen. Emílio Garrastazú Médici was elected President.

1970s: Opposition to military rule intensified.

January 1974: Gen. Ernesto Geisel was elected President.

November 1974: Legislative elections were held; the electoral system made a government majority inevitable.

October 1978: Gen. João Baptista da Oliveira Figueiredo was elected President. In congressional elections the following month, the MDB made substantial gains.

1979: Legislation to end the government-controlled two-party system was approved.

November 1982: The pro-Government Partido Democrático Social (PDS—as ARENA had been reconstituted following the end of the two-party system) lost significant numbers of seats in elections to the Chamber of Deputies, state governorships and municipal councils.

July 1984: Liberal members of the PDS and the Partido do Movimento Democrático Brasileiro (PMDB) formed an alliance which subsequently became known as the Partido Frente Liberal (PFL).

January 1985: Tancredo Neves of the PFL was elected President, gaining 480 of the 686 votes in the electoral college. He died before he could be inaugurated and the vice-presidential candidate, José Sarney, assumed the office on a temporary basis.

May 1985: The Constitution was revised to enable the direct election of the head of state by universal suffrage.

March 1986: The cruzeiro was replaced as the country's currency unit by the cruzado, equal to 1,000 cruzeiros.

15 November 1986: The PMDB won an absolute majority in elections to Congress, which was to function as a constituent assembly, and was inaugurated in February 1987.

June 1988: The approval of a presidential mandate of five years by the Congress prompted the resignation of several members of the PMDB. They subsequently formed a new movement, the Partido da Social Democracia Brasileira (PSDB).

6 October 1988: The new Constitution was promulgated; it provided for the transfer of many presidential powers to the legislature. Censorship and legislation which enabled dissidents to be detained without trial were also abolished.

January 1989: The cruzado was replaced by the new cruzado, equal to 1,000 cruzados and initially at par with the US dollar.

May 1989: The USA named Brazil, India and Japan as unfair trading partners, on the basis of Section 301 (the so-called 'Super 301' provision) of the US Omnibus Trade and Competitiveness Act of August 1988.

17 December 1989: Fernando Collor de Mello of the Partido de Reconstrução Nacional (PRN), won the presidential election, gaining 53% of the votes cast. Collor de Mello, previously the Governor of the State of Algoas, had been a relatively minor political figure during the 1980s, and indeed until the start of his presidential campaign; however, it was this quality of being an 'outsider' which won him the support of a large section of the national media and increased his popularity among the electorate.

March 1990: President Collor de Mello announced a programme of economic reform, aimed at reducing public expenditure and liberalizing the economy, and known as the 'New Brazil' plan. As part of the programme, the new cruzado was replaced at par by the cruzeiro.

3 October 1990: Elections to the Chamber of Deputies resulted in the PMDB suffering significant losses, winning 109 of the 503 seats, compared with 259 seats at the previous election; the PFL won 92 seats (a net loss of 13), the PDT 46 (an increase of 12) and the PRN 41 in its first election.

25 November 1990: The PMDB suffered a number of defeats in gubernatorial elections, including those in São Paulo, Rio Grande do Sul, Rio de Janeiro and Espírito Santo.

26 March 1991: The Treaty of Asunción was signed by Argentina, Brazil, Paraguay and Uruguay, committing the four countries to the establishment of a common market, to be known in Portuguese as the Mercado Comum do Sul and in Spanish as the Mercado Común del Sur (Mercosul/Mercosur—the Southern Common Market).

May 1991: Following the failure of a second series of economic reforms (the country had experienced a recession in 1990 followed by a resumption of inflation in 1991), the Government's economic team, including the Minister of the Economy, Zélia Cardoso de Mello, was forced to resign.

May 1992: President Collor de Mello was accused of corruption in association with his 1989 electoral-campaign treasurer, Paulo César Farias. The President's brother, Pedro Collor de Mello, was one of his principal accusers.

June 1992: The UN Conference on Environment and Development (popularly known as the 'Earth Summit') was held in Rio de Janeiro. The Brazilian Government had been accused of long-standing abuse of the environment, by permitting tropical rainforest to be logged and cleared for farming or gold-prospecting. In addition, indigenous-rights groups claimed that the development of industry had brought pollution and disease to indigenous tribes previously isolated from development. It was estimated that one group, the Yanomami, had lost 10%–15% of its total population since the beginning of gold-prospecting in their region.

29 September 1992: In response to a report by an inquiry into the allegations surrounding President Collor de Mello, the Chamber of Deputies voted to impeach him; the Cabinet subsequently resigned, and the President was suspended from office. The Vice-President, Itamar Cantiero Franco, assumed office as Acting President.

29 December 1992: Immediately following the Senate's decision to proceed with impeachment, President Collor de Mello announced his resignation. Franco was confirmed as President.

January 1993: The Supreme Court announced that the former President, Collor de Mello, was to stand trial, as an ordinary citizen, on charges of passive corruption and criminal association.

April 1993: In a national referendum, the electorate rejected the both the re-introduction of a parliamentary system of government and the restoration of a monarchy.

April 1993: The PDT and later the Partido Socialista Brasileiro (PSB), withdrew from the ruling coalition, in response to President Franco's decision to proceed with the privatization of the National Steel Company.

May 1993: Fernando Henrique Cardoso was appointed Minister of Finance; during his tenure he attempted to introduce major structural reforms to the economy; he had greatest success in altering the financial relationship between the federal and state governments.

August 1993: A new currency, the cruzeiro real, was introduced. It was equivalent in value to 1,000 cruzeiros.

August 1993: Following a territorial dispute with gold-prospectors, in which 73 Yanomami were killed, the Government announced the creation of a new post of Minister with Special Responsibility for the Brazilian Amazon.

March 1994: Cardoso, who had proved popular among the business community, resigned as Minister of Finance in order to be eligible for election to the presidency (the election was scheduled for October and candidates were required to have held no government post in the previous six months).

May 1994: The Constitution was amended to reduce the presidential term of office to four years.

July 1994: The cruzeiro real was replaced as the Brazilian currency unit by the real, which was at par with the US dollar at its introduction, and thus equivalent to 2,750 cruzeiros reais.

3 October 1994: Cardoso, who was supported by the PFL, the PTB and the PL, won the presidential election, gaining 54.3% of votes cast, defeating Luis Inácio 'Lula' Da Silva, of the Partido dos Trabalhadores (PT), who received 27.0% of the ballot. Cardoso attempted to reform the country's social-security and administrative systems, and announced his intention to privatize certain state-owned businesses.

December 1994: The Supreme Federal Tribunal voted to acquit the former President, Collor de Mello, of the corruption charges against him.

31 December 1994: Mercosul/Mercosur was established; its provisions came into effect the following day.

May 1995: The Central Unica dos Trabalhadores organized a general strike in protest at the privatization of major industries; the action was ended by the intervention of the armed forces and the Government proceeded with the privatization of shipping routes, natural-gas production and telecommunications.

September 1995: The Chamber of Deputies approved legislation by which the state assumed responsibility for the deaths of more than 100 left-wing politicians and activists during the 1960s and 1970s.

January 1996: The President enacted legislation which regulated the demarcation of indigenous lands; this prompted protests from indigenous-rights groups, who interpreted this as an infringement of their existing rights.

March 1996: It was reported that the Government had withheld details of a fraud involving US $5,000m. at the Banco Nacional ten years previously.

March 1996: An internationally funded programme to protect the rain forests and the indigenous groups inhabiting them was approved.

May 1996: A National Plan for Human Rights was introduced, containing increased rights to protection for workers.

September 1996: Three right-wing parties merged to create the Partido Progressista Brasileiro (PPB).

January 1997: The Chamber of Deputies approved legislation to allow an incumbent President, Governor or Mayor to be re-elected.

July 1997: A report of an investigation by the Senate into a financial scandal involving fraudulent bond issues implicated 20 prominent politicians and senior officials, in addition to 161 financial institutions.

August 1997: During conflict between riot police and illegal occupants of public land in Brasília, 49 were arrested and 20 injured. The perceived lack of progress in enacting land-reform legislation in Brazil angered landless farmers.

January 1998: Fernando Collor De Mello was exonerated of all charges of corruption.

August 1998: During a period of uncertainty on international financial markets investors began to remove capital from Brazil, causing a decline in the country's international reserves and damaging its ability to defend the real.

late October 1998: President Cardoso's administration announced a three-year plan of fiscal reform, with the aim of allowing Brazil to gain access to funds provided by the International Monetary Fund (IMF) and other donor organizations.

4 October 1998: Cardoso was re-elected to the presidency, receiving 53.1% of votes cast, compared with 31.7% of the ballot obtained by Lula da Silva. However, a high rate of abstention was reported. In concurrent legislative elections, the PFL secured 106 of the 513 seats in the Chamber of Deputies, the PSDB winning 99, the PMDB 82, the PPB 60 and the PT 58.

November 1998: The Minister of Communications and the Foreign Trade Secretary (who were brothers) and the President of the National Development Bank were forced to resign following the revelation, through the unlawful recording of telephone conversations, that they had attempted to influence the sale of the state-owned telecommunications company earlier in the year.

2 December 1998: The Chamber of Deputies rejected elements of the Government's fiscal-reform programme—this caused a further lack of confidence in Brazilian investments and a decline in foreign-exchange reserves.

13 January 1999: The fixed exchange rate for the real was abandoned, following a state's default of its debt to the federal government. The real subsequently depreciated greatly against major currencies, rendering the Government's plans for economic restructuring moribund. The Governor of the Central Bank, Gustavo Franco, resigned and was replaced by Francisco Lopes.

1 February 1999: Lopes was dismissed; in April he was arrested in connection with his refusal to give evidence to an inquiry into corruption at the Central Bank. Armínio Fraga Neto was appointed to succeed Lopes.

March 1999: The PTB announced its withdrawal from the ruling coalition in response to the dismissal of the Minister of the Budget and Administration, Paulo Paiva.

July 1999: President Cardoso reorganized the Government, appointing Clovis Carvalho to the post of Minister of Development, Industry and Trade.

August 1999: Anti-Government demonstrations organized by the PT and the trade unions took place in Brasília in protest at President Cardoso's economic and social policies.

September 1999: Carvalho, who had publicly disagreed with President Cardoso's policies of economic austerity, was dismissed; he was replaced by Alcides Tápias.

October 1999: A congressional commission of investigation into organized crime revealed that a criminal network had influence at the highest levels of politics, business and the judiciary; a new anti-corruption force was established the following month.

January 2000: The Minister of Defence, Elicio Alvares, was dismissed; he was replaced by Geraldo Magela Quintão.

February 2000: The PSDB announced that it was to form a congressional alliance with the PTB, thereby marginalizing the PFL.

March 2000: The Mayor of São Paulo, Celso Pitta, was accused by his wife of bribing members of the city council to defeat a motion proposing his dismissal in 1997. Following further allegations of corruption made against him, Pitta was dismissed later in the month, although he was permitted to remain in office pending an appeal.

May 2000: Some 30,000 members of the landless peasant movement, Movimento dos Trabalhadores Rurais Sem-Terra (MST), occupied government buildings, including the Ministry of Agrarian Reform. Following negotiations with government representatives, the MST agreed to leave the buildings and talks on their grievances regarding the land-reform process were resumed.

June 2000: A federal court upheld Pitta's appeal against his dismissal and he was allowed to remain in office.

July 2000: A number of associates of President Cardoso were implicated in a scandal regarding unlawful payments made in connection with the construction of a courthouse in São Paulo—one, a Senator, was expelled from Congress.

August 2000: The PTB dissolved the congressional alliance it had formed with the PSDB, citing its concerns over the persistent allegations of corruption among PSDB officials and the Government's consequent decline in popularity.

October 2000: In land-reform legislation, the Government offered loans to those already settled on their own land.

November 2000: A further report on drugs-trafficking and crime implicated a number of senior political figures.

February 2001: The Ministers of Social Security and Assistance, and of Mines and Energy, were dismissed following disagreements on policy within the Government.

31 March 2001: The President of the USA, George W. Bush, met with President Cardoso for discussions on the creation of a new regional trade zone.

20 April 2001: The Government announced the creation of 14 new reserves for indigenous peoples, which would encompass a total area of 2m. ha.

May 2001: The Minister of National Integration, Fernando Bezzora, resigned, following allegations of corruption made against him.

30 June 2001: The former head of the police force, Col Ubiratan Guimarães, was convicted of the murder of 102 inmates in a riot in a prison in São Paulo in October 1992.

1 September 2001: The Government announced it had reached an agreement with the Swiss pharmaceutical company, Roche, reducing the price of a drug used in the treatment of Acquired Immunodeficiency Syndrome (AIDS). The Government had previously announced its intention to permit the manufacture of the drug within the country in contravention of Roche's patent if the company did not reduce its price. In August the USA had abandoned a threat to take legal action against Brazil if it permitted the manufacture of generic versions of patented anti-AIDS drugs.

16 September 2001: The Government and the IMF agreed a series of loans to be made to Brazil during 2001–02, to the value of US $15,600m. The IMF stated that Brazil's economic progress since its 1998 crisis had been admirable; however, uncertainty on global financial markets following the terrorist attacks on the USA earlier in the month had caused a significant depreciation in the value of the real and was expected to precipitate a decline in investment in the country.

6 November 2001: The trial of four men (all members of prominent families) accused of the murder of an indigenous leader in Brasília in 1997 began in the capital. Indigenous-rights groups campaigned outside the court, claiming that the accused, who admitted causing the death, but claimed it was an accident, would receive lenient treatment for the offence.

Canada

c. 30,000 BC: During the late Pleistocene Epoch of the Quaternary Period, the last Ice Age exposed a land bridge, subsequently submerged beneath the Bering Sea, across which Asian peoples migrated from the north-eastern part of what is now Siberia to what is now Alaska. They gradually spread throughout the Americas and were the first known inhabitants of the continent. Sparse archaeological evidence indicates that these peoples were probably nomadic hunter-gatherers who lived in groups of about 100 individuals, used stone tools and tanned reindeer or caribou skins.

22,000 BC: Radiocarbon dating indicates that bone tools found in the Yukon Territory of Canada were fashioned from animals that had lived at this time.

***c.* 2000 BC:** A burial mound found at Port-au-Choix, on Newfoundland, is believed to have been constructed at this time.

AD 986: Norseman Bjarni Herjólfsson, while attempting to sail from Iceland to Greenland, was blown off course by a storm and ended up on an unknown shore now believed to have been on Canada's east coast. If so, he and his party may have been the first Europeans to visit the American continent.

***c.* 1000:** The Icelandic navigator Leif Ericsson, who had purchased Herjólfsson's ship, travelled down the east coast of what is now Canada. He visited places he called Helluland ('stone land', thought to be Baffin Island), Markland ('forest land', thought to be Labrador) and Vinland ('vine land', after the many native North American grapes he saw growing on an island thought to be Newfoundland); whether he sailed as far south as Nova Scotia and New England is debated.

1004–05: According to Icelandic *sagas*, during this winter native inhabitants of Vinland killed Thorvald Ericsson, who was a brother of Leif Ericsson.

early 11th century: Vikings, probably including Leif Ericsson himself and other members of his family, established at L'Anse-aux-Meadows, on the northern tip of Newfoundland, a temporary settlement of wood-framed peat houses of a type found in Greenland. The site, unearthed by Norwegian archaeologists in the 1960s, matches the original description of Vinland. Evidence of iron-working was also found, as were artefacts of iron, bronze, wood, stone and bone.

1497–98: The Venetian navigator Giovanni Caboto, sailing under the name John Cabot in the service of the English monarch, King Henry VII, explored along North America's east coast between Labrador and the inlet now known as Chesapeake Bay, in the north-eastern USA.

1500: The Portuguese navigator Gaspar Corte-Real reached the coast of Labrador and the Avalon Peninsula of south-eastern Newfoundland. He embarked on a second voyage the following year, but failed to return.

April 1534: The French navigator Jacques Cartier sailed from St-Malo with two ships and reached Newfoundland 20 days later. He planted a cross on the Gaspé Peninsula, now in the south of Québec, and claimed the territory for France. He then circumnavigated the Gulf of St Lawrence and explored the territories now known as New Brunswick and Prince Edward Island before returning to France.

1535: Cartier embarked on his second voyage, this time sailing up the St Lawrence River, which he named, to the indigenous settlement of Stadacona (now Québec City) and then to the indigenous settlement of Hochelaga. At the latter site, Cartier climbed a hill and named it Mont Réal.

1541–42: On his third voyage, Cartier sailed up the St Lawrence River as far as Lachine Rapids and made an unsuccessful attempt to establish a French settlement there.

1576–78: The English navigator and explorer Sir Martin Frobisher made three voyages to Canada in search of the 'North-West Passage', a sea route from the Atlantic to the Pacific across the northern coast of the American continent; this route was considered to be of great importance by Europeans seeking to facilitate trade with the littoral countries of the Pacific, although it would be more than three centuries before it was successfully navigated.

1583: The English navigator Sir Humphrey Gilbert named the island of Newfoundland and claimed it for the English Crown.

1600: The indigenous population of what is now Canada is estimated to have exceeded 250,000 inhabitants at this time. They comprised Eastern Woodland peoples—nine farming tribes including the Iroquois Confederacy of southern Ontario and the Algonquin of western Québec; Great Plains peoples, who were nomadic livestock-rearers and comprised four main tribes including the Blackfoot of south-eastern Alberta and the Assiniboine of southern Saskatchewan; Plateau peoples, 11 tribes in eastern British Columbia; Subarctic peoples, who hunted caribou and moose and comprised 31 tribes sparsely populating subarctic Canada from Alaska to the Atlantic, the best known being the Cree; North-West Pacific Coast peoples, who inhabited a thin strip all along British Columbia's Pacific coast and several offshore islands and comprised 14 tribes including the Tlingit and the Kwakiutl; and Arctic peoples, including the Inuit who, after about 2000 BC, had migrated from Alaska across Canada to Greenland and hunted seal, caribou and whale.

17th century: France established settlements around Placentia Bay, on the south coast of Newfoundland, and subsequently on the west coast.

1603: France, under King Henry IV of Navarre, granted a monopoly over New France to the French Huguenot (Protestant) explorer Pierre du Guast, Sieur de Monts who, in the same year, sailed there with the explorer Samuel de Champlain.

1604: Champlain and de Monts made another voyage to North America with a small party of French settlers who founded a colony at the mouth of the Saint-Croix River, which today forms the border between the Canadian province of New Brunswick and the US state of Maine. The surrounding area, comprising present-day New Brunswick, Nova Scotia and Prince Edward Island, became known as Acadia.

1605: The Saint-Croix River colonists abandoned their site and moved to what is now south-western Nova Scotia, where they founded Port Royal, the first permanent European settlement in North America. Port Royal was subsequently renamed Annapolis Royal, after the British Queen, Anne.

1608: King Henry IV appointed Champlain as Lieutenant-Governor of New France, whereupon he made his fourth voyage from France to Canada and founded Québec City; it subsequently became the administrative centre of New France. Champlain (who subsequently became popularly known as the 'father of New France') forged alliances with the Algonquin and Huron tribes that endured for more than two centuries.

1610: British settlers founded the colony of Saint John's, on the Avalon peninsula.

1610: The English navigator Henry Hudson made a fourth voyage in search of the North-West Passage. Under the patronage of an English company he sailed his ship *Discovery* into the bay that now bears his name and explored its eastern shore before becoming icebound at the onset of winter. Upon the thaw of the ice, permitting a return to the Atlantic, in June 1611, the crew mutinied and abandoned Hudson in the bay.

1611: Champlain founded a trading post at Hochelaga.

1621: Citing the 1497–98 voyage of John Cabot in the service of the English crown, Scottish-born King James I decided to press Britain's claim to Acadia by granting proprietorship over it to the Scottish statesman and poet, Sir William Alexander, with a view to recruiting British colonists to settle there. He gave named part of the territory Nova Scotia, meaning 'New Scotland'. Rivalry between Great Britain and France for primacy in Canada began.

1627: The French cardinal and statesman Armand Jean du Plessis, Duc de Richelieu, who was Chief Adviser to the King Louis XIII, established the Company of One Hundred Associates, for the purpose of establishing French Roman Catholic colonies in what was officially known as New France. This name would remain in use until 1763.

1629: British forces occupied Québec City, the capital of New France.

1632: Great Britain yielded control of New France and Acadia to France.

1642: French settlers founded Ville Marie on an island in the St Lawrence River, near the hill which Cartier had named Mont Réal; the settlement would become the city of Montréal.

1645: The late Cardinal Richelieu's Company of One Hundred Associates appointed the Community of Habitants as its colonial agents.

1648–49: The Iroquois confederacy, known since the 16th century as the Five Nations, attacked the Huron tribe and then threatened the French settlers, who were allies of the Huron.

1663: Jean-Baptiste Colbert, adviser to King Louis XIV, placed New France under the direct control of the French crown.

1666: The population of New France had grown to about 2,000 inhabitants of European, mainly French, descent. France granted large proprietorships to a select few men, called *seigneurs*, who were deemed to have sufficient wealth and ability to attract countrymen, called *habitants*, to emigrate to the 'New World' (as the Americas had become known), settle on their land and make it productive, usually as tenant farmers.

1670: Britain's restored King, Charles II granted a Royal Charter for the Hudson's Bay Company to a consortium of European aristocrats. The company was given control over territory on and to the west of Hudson Bay.

1689–97: King William's War, named after Britain's Dutch-born King William III, was the name given to the North American theatre of the conflict known variously in Europe as the War of the Grand Alliance, the War of the League of Augsburg and the War of the English Succession, in which Great Britain, the Netherlands, Spain and Austria formed the 'grand alliance' to fight against France. War erupted in America after a series of English raids on French settlements, when the Governor of New France, Louis de Buade, Count of Frontenac, retaliated by attacking and burning the British colonial town of Schenectady, New York and threatening Boston and New York City. French pirates based in Acadia disrupted British ships approaching American ports. English colonial troops then captured the French colony of Port Royal.

20 September 1697: The Treaty of Ryswick, which concluded the War of the Grand Alliance, returned Port Royal to France.

1704: British colonial troops retaliated to a French attack on a British settlement by attempting to recapture Port Royal. The two European powers were opposed in the War of the Spanish Succession.

1710: British colonial troops, aided by the British Royal Navy, occupied the French colony of Acadia, but their subsequent assault on Québec City and Montréal failed.

1713: The Treaty of Utrecht, which formalized the end of the War of the Spanish Succession, awarded Newfoundland, continental Acadia, Hudson Bay and the islands of St Pierre and Miquelon to Great Britain.

1744–48: During the War of the Austrian Succession in Europe, fighting erupted in North America when French troops seized a British fort at Canso in Nova Scotia, imprisoned its occupants and transported them to the French fort at Louisbourg on Cape Breton Island.

April 1745: A British colonial force from Boston besieged Louisbourg; the fort eventually capitulated on 15 June.

18 October 1748: The Treaty of Aix-la-Chapelle (now Aachen, Germany) concluded the War of the Spanish Succession; under its terms, Louisbourg was returned to the French.

1749: British settlers under the leadership of Col Edward Cornwallis founded Halifax, the future capital of the province of Nova Scotia.

1753: The French began erecting a line of fortresses extending from Lake Erie south to the fork of the Ohio River, in what today is the USA. The completion of the southernmost of the fortresses, Fort Duquesne, at what is now Pittsburgh in Pennsylvania, USA, directly challenged Britain's Virginia-based Ohio Company, which had also sought to establish a fortress in the same position.

1754: The Scottish-born Lieutenant-Governor of the Virginia colony, Robert Dinwiddie sent a military force under the command of Col George Washington, the future first President of the USA, to challenge the French; the ensuing conflict became known as the French and Indian War, more commonly known (from 1756) as the American theatre of the Seven Years' War, in which Great Britain and its allies opposed France.

1755: British colonial authorities began requiring that all residents of Acadia, including those of French descent, swear allegiance to the British Crown. Any who refused had their lands confiscated and were forcibly deported to other colonies of the British Empire.

1758: British colonial troops again captured Louisbourg, destroyed Fort Frontenac on Lake Ontario and seized Fort Duquesne.

1758: British North America's first elected legislative assembly met at Halifax, Nova Scotia.

13 September 1759: In a decisive victory, British forces defeated the French main army at the Battle of the Plains of Abraham, overlooking Québec City.

8 September 1760: Montréal fell to the British.

10 February 1763: The Treaty of Paris, which concluded the Seven Years' War, awarded most of the French Empire in North America to Great Britain, which immediately ceded the islands of St Pierre and Miquelon to France.

1763: Having gained sovereignty throughout present-day Canada, the British Crown determined the boundaries of Québec and established a new government there; Cape Breton Island was annexed to Nova Scotia and Labrador was united with Newfoundland.

1774: The British Parliament passed the Québec Act, extending the colony's boundaries to include Labrador, guaranteeing religious and linguistic freedom and granting French-speaking Québecois equal status with English-speakers.

1775–83: During the War of American Independence in which the 13 colonies to the south of Canada defeated the British and won the right to become the independent United States of America, an estimated 50,000 citizens of the 13 original colonies opposed independence, refused to swear a loyalty oath to the new government and fled north to Canada, where they became known as United Empire Loyalists.

1776: French traders in Montréal founded the North West Company.

1778: The British navigator James Cook attempted to sail the Northwest Passage, this time from the Pacific side. He failed, although he was able to chart the whole of Canada's Pacific coast, including Vancouver Island, before winter forced him to retreat.

1783: The Treaty of Paris that followed the end of the War of American Independence defined in Article II the border between the USA and Canada (although confusion as to the map used resulted in the border being disputed). Further loyalists left the USA and emigrated to the western part of the Canadian colony of Nova Scotia, where they received generous land grants from the British government.

1784: The fragmentation of Acadia began when New Brunswick was created a separate colony from Nova Scotia; the move had been prompted by the recent large influx into New Brunswick of loyalist settlers from the USA.

1786: St John's Island (later renamed Prince Edward Island) was created a separate colony from Nova Scotia.

1791: The Spanish navigator José Maria Narváez visited the site of present-day Vancouver and claimed it for Spain.

1791: The Constitutional Act established limited self-government in Canada; the colony was divided into Upper Canada (now the southern part of Ontario province) and Lower Canada (now the southern part of Québec province), each with its own system of government.

1792: Spain ceded the territories claimed by Narváez to the British. In the same year, the British navigator George Vancouver, surveyed the area during an attempt to locate the North-West Passage.

1793: The British colonial government founded the town of York on the north-western shore of Lake Ontario, to serve as a purpose-built capital of Upper Canada; it subsequently grew into Toronto.

1795: Fort Edmonton was founded on the North Saskatchewan River as a fur-trading post.

1809: Labrador was transferred from the Québec colony to the Newfoundland colony.

1812–15: US forces attempted to end British influence in North America by expelling British troops from Canada, but were unsuccessful.

30 May 1814: The islands of St Pierre and Miquelon were permanently restored to France.

1815–18: The Métis (Canadians of mixed French and indigenous descent) rebelled in the Red River area of what is now Manitoba province.

1818: The Convention of 1818 partially ended disagreement between the USA and the United Kingdom and defined the border between the USA and Canada as the 49th parallel (49°N) between the Great Lakes and the Rockies.

1821: The Hudson's Bay Company merged with the Northwest Company under the name of the former.

1825: A treaty between the British and Russian Empires demarcated the land border between Alaska, then Russian, and Canada.

1832: The United Kingdom granted the right of representative government to the British colony of Newfoundland.

6 March 1834: York was renamed Toronto, after a word of the Huron language meaning 'place of abundance' or 'meeting place'.

1837: Rebellions in Montréal and Toronto sought a wider measure of self-government for Francophone Canadians.

1840: The Act of Union united Upper and Lower Canada, renaming them Canada West and Canada East, respectively.

1842: The Webster-Ashburton Treaty permanently fixed the eastern part of the US–Canadian border, from Maine and New Brunswick in the east to Minnesota and western Ontario in the west.

1846: The Oregon Treaty permanently settled the remainder of the US-Canadian border, extending the 49th-parallel border westward to the Pacific Ocean (although Vancouver Island, which traversed the line, was specified as entirely Canadian).

1848: The joint 'reform ministry'—Canada's first real cabinet—was formed.

1849: Vancouver Island became a British colony.

1855: The Newfoundland colony achieved full self-government within the British Empire; a bicameral Parliament was established.

1855: The settlement of Bytown was renamed Ottawa, after the indigenous Outaouac people.

1857: Ottawa was selected as capital of the United Province of Canada.

1858: Colonial status was granted to British Columbia, following the discovery of gold in the Fraser and Thompson Valleys and the immigration of large numbers of gold-prospectors.

1864: A conference in Charlottetown, Prince Edward Island, discussed unification of the Maritime (Atlantic) colonies; Canada obtained permission to attend as an observer. This conference led to the incorporation of the future Maritime provinces into Canada, although initially Prince Edward Island was not one of them. The Québec Conference, held less than six weeks later, drafted an agreement for the confederation of all the British colonies north of the USA.

1866: Vancouver Island united with British Columbia.

1 July 1867: The British North America Act (known in Canada as the 1867 Act of Confederation, and later known as the first Constitution Act) formed the Dominion of Canada, comprising initially the provinces of New Brunswick, Nova Scotia, Ontario and Québec. Sir John Alexander MacDonald, a Scottish-born Conservative from Kingston, Ontario, took office as the first federal Prime Minister; he would serve for the next six years. The first Parliament convened at Ottawa.

1867: In a referendum, Newfoundland's voters opposed joining the Dominion of Canada.

1869: The Hudson's Bay Company sold extensive lands, including Rupert's Land and other western and north-western territories, to the Canadian government. The Métis, fearing encroachment of their freedoms, opposed the sale.

1869: Louis David Riel, born in Manitoba of mixed French, indigenous and Irish descent, led the Métis in the Northwest Rebellion. They seized Fort Garry at Winnipeg and proclaimed their secession from Canada, but the revolt was swiftly suppressed.

1870: The Canadian government formally designated the lands it had purchased from the Hudson's Bay Company as Manitoba and the Northwest Territories; both entities joined the Dominion of Canada, and Manitoba became Canada's fifth province.

1871: British Columbia entered the Dominion, becoming Canada's sixth province, on the condition that Canada build a trans-continental railway linking the west coast to the east.

1873: Canada founded the Northwest Mounted Police (later renamed the Royal Canadian Mounted Police) to maintain order in the vast, remote Northwest Territories.

1873: Prince Edward Island entered the Dominion and became Canada's seventh province.

1873: Alexander Mackenzie, also Scottish-born and from Kingston, Ontario, led the Liberals to victory in a general election and became Canada's second Prime Minister.

1878: The Conservatives returned to power, and Sir John MacDonald again became Prime Minister.

1880: An Imperial Order-in-Council in the United Kingdom formally annexed to Canada all British possessions in North America north of the USA, excepting Newfoundland but including the Arctic islands.

1885: The completion of the Canadian Pacific Railway linked British Columbia to Montréal.

1885: Riel returned from exile in the USA and led the Métis in another rebellion, this time in Saskatchewan. Government forces again suppressed the revolt and took Riel into custody. He was subsequently convicted of treason and executed.

April 1886: Vancouver was incorporated as a city.

1888: The self-governing Parliament in the British colony of Newfoundland enacted legislation prohibiting the sale of bait to French fishermen who resided on the island, escalating tension between the two groups of residents on the island.

1891: Sir John Joseph Caldwell Abbott, a Conservative from St Andrews, Québec, led his party to victory and became the first Canadian Prime Minister to have been born in North America.

1892: Abbott resigned owing to ill health; Sir John Sparrow David Thompson, born in Halifax, Nova Scotia, replaced him.

1894: Thompson died in office; Sir Mackenzie Bowell, an English-born Conservative, became Prime Minister.

1895: In a referendum, the voters of Newfoundland again rejected membership of the Dominion of Canada.

April 1896: Bowell resigned amid widespread criticism of his policies and his leadership; his successor, Sir Charles Tupper, was unable to improve the Conservatives' popularity.

July 1896: The Liberal Party, led by Sir Wilfrid Laurier, won the general election. Laurier became the first Prime Minister of French descent.

17 August 1896: Alluvial gold was discovered in Bonanza Creek, a tributary of the Klondike River, in the west of the Northwest Territories.

mid-1896: News of the Klondike gold strike emerged, leading to a 'gold rush' in which more than 30,000 prospectors entered the area by the end of 1897.

13 June 1898: Yukon Territory was created from the Northwest Territories, in the region surrounding the areas where gold-prospecting had brought immigration.

1903: The border between British Columbia and Alaska, by now a US possession after its purchase from Russia in 1867, was fixed by tribunal, allowing a narrow strip of Alaskan territory to occupy more than 900 km of British Columbia's Pacific coastline.

1903–06: After more than 300 years of failed attempts, the Norwegian polar explorer, Roald Amundsen, led the first expedition ever to navigate the North-West Passage successfully. He and his crew had to endure three icebound winters to achieve it.

1904: Under an Anglo-French agreement, France gave up its claim over any part of Newfoundland in exchange for certain territories in Africa. The self-governing British colony of Newfoundland now encompassed the entire island.

1905: The Northwest Territories Act established new boundaries in the region and created the provinces of Alberta and Saskatchewan, which joined the Dominion, becoming Canada's eighth and ninth provinces.

1909: The Boundary Waters Treaty established the International Joint Commission to settle disputes between the USA and Canada over territorial waters in lakes and rivers along their mutual border.

1911: The Conservatives, led by Sir Robert Laird Borden, returned to power.

1914: Canada supported the United Kingdom when it declared war on Germany, and the Dominion entered the First World War.

1917: Following the introduction of conscription, more than 500,000 Canadian troops had departed for Europe to fight in the First World War. The issue of conscription divided Anglophone and Francophone Canadians and caused the Conservatives' popularity to decline sharply. In a general election, the Conservatives lost their majority in the House of Commons, although Borden was able to continue as Prime Minister by forming a 'Unionist' coalition Government.

1918: Riots erupted in Québec City, in protest against conscription.

1919: Canada participated in the formation of the League of Nations and became a charter member in its own right.

10 July 1920: Borden resigned on health grounds. Arthur Meighen succeeded him as leader of the Unionist coalition and as Prime Minister; he was the first Prime Minister to have been born in the independent Dominion of Canada.

December 1921: The Unionist coalition dissolved and the Liberals formed a Government, led by William Lyon Mackenzie King.

1924: Farmers dissatisfied with the monopolistic power held by the railway and the grain-elevator owners founded the Saskatchewan Wheat Pool in Regina.

1925: The Conservatives emerged from a general election as the largest parliamentary grouping, but did not hold a majority in the chamber. The Liberals were able to form an alliance with the Progressive Party to remain in office, again led by King.

June 1926: King's government fell, and the Governor-General (the local representative of the head of state, the British monarch), asked Meighen to form a new administration. However, the minority Government was unable to command the support of the House of Commons, and a new general election was organized.

September 1926: The Liberals won 116 parliamentary seats in the general election and formed a new Government, led by King.

1927: The British Privy Council settled the long-standing dispute over the location of the border between Québec and Labrador which, as part of Newfoundland, remained a British possession.

1930: Richard Bedford Bennett, later the first Viscount Bennett, led the Conservatives to victory in a general election and became Prime Minister.

11 December 1931: The United Kingdom guaranteed British recognition of the autonomy and equal status of each of the Dominions of the Commonwealth, proclaiming them to be 'partner nations' to the United Kingdom and to be 'equal in status, in no way subordinate to each other'.

16 February 1934: Newfoundland, the economy of which had been particularly severely damaged by the global economic depression of the late 1920s and early 1930s, lost the self-governing status it had enjoyed for 79 years, by an Act of the British Parliament. Responsible governing authority was vested instead in a seven-member commission comprising the Governor, three British nationals and three Newfoundlanders.

1935: A Liberal victory in the Canadian general election returned King to the office of Prime Minister.

1939: Canada issued its own declaration of war on Germany and entered the Second World War.

26 March 1940: In a general election, the Liberals, again led by King, were returned to office with a large parliamentary majority.

1942: The Conservative Party was renamed the Progressive Conservative Party (PCP).

1945: Following the end of the Second World War, Canada became a charter member of the UN.

11 June 1945: King's Liberal Party retained its parliamentary majority after a general election.

22 July 1948: In a referendum, Newfoundland voted by 78,323 votes to 71,334 in favour of unification with Canada.

1949: The second British North America or Constitution Act enabled the Canadian Parliament to enact constitutional amendments in matters affecting the federal government.

31 March 1949: Newfoundland became Canada's 10th province.

27 June 1949: Louis Stephen Saint-Laurent led the Liberals to victory in a general election, following the retirement of King.

1952: Vincent Massey became Canada's first Canadian Governor-General.

10 August 1953: In a general election, the Liberals won 170 of the 244 seats in the House of Commons and were returned to office, led by Saint-Laurent.

1954: Construction began on the St Lawrence Seaway, a joint Canadian-US project to enable ocean-going vessels to reach the Great Lakes between May and November.

10 June 1957: The Liberal Party lost its majority in a general election. John Diefenbaker, the leader of the PCP which, with 112 of the 242 seats was the largest single party in the House of Commons, formed a minority PCP administration.

1958: The USA and Canada together founded the North American Aerospace Defense Command (NORAD).

31 March 1958: In a general election, the PCP won 208 of the 264 seats in the enlarged House of Commons, the Liberals winning only 48 and the Co-operative Commonwealth Federation (CCF) the remaining eight. Diefenbaker resumed the premiership.

25 April 1959: The St Lawrence Seaway came into operation, accepting its first ocean-going vessels.

1960: The Canadian Bill of Rights gained royal assent and was promulgated.

18 June 1962: In a general election, the PCP won 116 seats, the Liberals 99, the Social Credit Party 30 and the National Democratic Party (NDP—a new movement formed by the merger of the CCF and a number of smaller parties) 19. Diefenbaker remained in power, leading a minority government.

1963: Amid increasing discontent among French-speakers, who claimed that the country's administration favoured English-speakers, a Royal Commission on Bilingualism and Biculturalism was appointed.

8 April 1963: A general election was held; the principal issue was Canada's acquisition of nuclear weapons from the USA, to which Diefenbaker was opposed. The Liberals won 129 seats in the House of Commons, the PCP 95, Social Credit 24 and the NDP 17. Lester Pearson formed a Government.

15 February 1965: The red-and-white maple-leaf flag became the official flag of Canada.

8 November 1965: In a general election, the Liberals won 131 seats, the PCP 97, the NDP 21 and Social Credit 14. Pearson returned to office.

1966: Gold production ended in the Yukon Territory.

1967: While on a state visit to Canada, the French President, Gen. Charles de Gaulle, demanded a 'free Québec' during a speech in Montréal. The Canadian Government took exception to his remarks and President de Gaulle returned to France ahead of schedule.

20 April 1968: After Pearson resigned as Prime Minister and as Leader of the Liberal Party, Pierre Elliott Trudeau, a French-speaking Liberal, was sworn in as Prime Minister.

25 June 1968: Trudeau led his party to the first overall majority in the House of Commons since 1962, winning 155 of the 264 available seats, compared with 72 won by the PCP, 22 by the NDP and 14 by Social Credit.

5 October 1970: The Front de Libération du Québec (FLQ) abducted the British Trade Commissioner, James Cross, in Montréal; the event precipitated a seven-week period of tension and unprecedented emergency police powers.

10 October 1970: FLQ members abducted Pierre Laporte, the Minister of Labor in the provincial Government of Québec.

13 October 1970: Canada established full diplomatic relations with the People's Republic of China; relations with Taiwan (the Republic of China) were terminated.

16 October 1970: Trudeau declared that Canada was in a state of 'apprehended insurrections' and invoked the War Measures Act of 1914, never before implemented in peacetime; the House of Commons approved the use of the Act by a vote of 190 to 16, with only the NDP opposing its introduction. This Act brought the armed forces onto the streets in Québec, allowed the detention of suspects for 21 days without specific charges and permitted the authorities to imprison suspects without bail for up to 90 days.

17 October 1970: Laporte was killed by his captors.

3 December 1970: After police surrounded a house, Cross was released unharmed in a negotiated deal which allowed his captors to be flown to Cuba and escape arrest. On the same day, the War Measures Act was withdrawn and was replaced by the Public Order (Temporary Measures) Act of 1970, with many of the same provisions.

January 1971: A number of FLQ activists were arrested and charged with Laporte's murder and/or abduction; those convicted were imprisoned.

30 April 1971: The Public Order (Temporary Measures) Act was suspended after five months.

June 1971: The Premier of Québec, Robert Bourassa of the Liberal Party, opted against approving a new constitutional charter, which had already been ratified by Trudeau and by all the other provinces' premiers.

18 October 1971: The premier of the USSR, Aleksei Kosygin, while on a nine-day visit to Canada, was attacked near the Parliament Buildings in Ottawa by a Hungarian anti-Communist activist but was unhurt; anti-Soviet demonstrators followed Kosygin throughout his visit.

30 October 1972: Trudeau's Liberal Party lost its majority in a general election, winning 109 seats in the House of Commons, compared with 107 won by the PCP and 31 by the NDP; Trudeau formed a coalition with the NDP and remained in office.

8 July 1974: Following the collapse of the governing coalition, a general election was held. The Liberals regained a majority in the legislature, winning 141 seats, compared with 95 won by the PCP and 16 by the NDP.

1976: The death penalty was abolished throughout Canada.

November 1976: The separatist Parti Québecois (PQ) came to power following provincial elections in Québec. The new premier, René Lévesque, advocated the full independence of Québec, issued draft legislation proclaiming an end to bilingualism in the French-speaking province and appointed an all-Francophone Cabinet. Schools throughout Québec were to conduct classes only in French.

January 1977: Trudeau, himself a Francophone Québecois, announced plans for an imminent referendum on the independence of Québec, in exchange for Lévesque's promise to resign if the electorate were to vote against secession. Thereafter, the PQ moderated its position, advocating phased increases in autonomy rather than immediate full independence.

15 February 1977: A new Citizenship Act came into force, and Canadians' status as 'British subjects' ended.

May 1977: The Berger Commission, yielding to pressure from indigenous groups living in the area, called for a 10-year delay in the start of construction of a natural-gas pipeline in the Northwest Territories.

22 May 1979: Trudeau's Liberal Party lost a general election held on the very last day allowable, and Trudeau resigned as party leader. Joseph Clark led the PCP to 136 parliamentary seats, compared with the Liberals' 114 and the NDP's 26, and formed a minority PCP administration.

November 1979: Trudeau resumed the leadership of the Liberal Party.

13 December 1979: The Canadian Supreme Court annulled a Québec provincial law proclaiming French to be the only official language of the province.

1980: A referendum in Québec rejected the PQ's proposals for secession from Canada (under the title of 'sovereignty-association') by almost 60% of those voting.

18 February 1980: Following a number of parliamentary defeats for Clark's minority Government, a general election was held; the Liberals won 147 of the 282 available seats, compared with 103 won by the PCP and 32 by the NDP, and returned to power; Trudeau again became Prime Minister.

November 1981: Nine of the 10 provinces (the dissenting province was Québec) and the federal government approved the 'patriation' of the Canadian Constitution, empowering the Canadian legislature to make amendments to it.

17 April 1982: The British Parliament approved the Canada Constitution Act ending the last formal statutory vestiges of Canada's colonial status by empowering Canada to amend its own Constitution.

February 1984: Trudeau announced that he was to resign as Prime Minister and as leader of the Liberal Party.

30 June 1984: John Napier Turner, who had been elected Liberal leader on 16 June, was sworn in as Prime Minister.

4 September 1984: Brian Mulroney led the PCP to a convincing victory in a general election, winning 211 seats in the House of Commons, the Liberals winning 40 and the NDP 30. Mulroney formed the first PCP majority Government since that formed after the general election of 1958.

1985: Lévesque retired; Pierre-Marc Johnson succeeded him as PQ Leader and as Premier of Québec.

3 June 1987: The Meech Lake Accord was finalized, recognizing Québec province as a 'distinct society' within the Canadian federation and granting the provinces new powers. It would, however, require ratification by the federal Parliament in Ottawa and by all 10 provincial legislatures by 23 June 1990.

1988: The province of Québec overruled the Canadian Supreme Court's earlier decision that a Québecois law requiring all public signs to be written only in French was unconstitutional.

21 November 1988: In a general election, the PCP won 169 of the 295 seats in the expanded House of Commons, the Liberals winning 83 and the NDP 43. Mulroney continued as Prime Minister.

1989: The Free Trade Agreement (FTA) between Canada and the USA came into effect.

1990: Armed confrontations between the civil authorities and militant indigenous-rights groups took place in Québec.

June 1990: Turner resigned as Leader of the Liberal Party. His successor was Joseph Jacques Jean Chrétien, a francophone from Québec.

23 June 1990: The Meech Lake Accord was abandoned when the period in which all provinces were required to have ratified it expired. The Newfoundland legislature had endorsed the Accord, then rescinded its endorsement, and Manitoba also opposed the agreement.

May 1992: A referendum in the Northwest Territories approved a proposal to divide the Territories and create a self-governing homeland for the Inuit people. To be called Nunavut, the homeland was to be created by 1999.

26 October 1992: The Charlottetown Agreement, a new federal-provincial constitutional compact designed to reassure Québec, was rejected in a national referendum. Only four of the 10 provinces approved the proposed arrangements, which were opposed by a number of the major political parties.

24 February 1993: After his popularity had declined sharply, Mulroney announced that he was to resign as soon as the PCP chose a new Leader.

13 June 1993: Avril Phaedra ('Kim') Campbell, the Minister of Defense was elected leader of the PCP, and thus Prime Minister, the first woman to hold the office.

25 October 1993: In a general election, the PCP suffered the heaviest defeat of an incumbent party in Canadian history, retaining only two seats in the 295-seat House of Commons. The Liberal Party won 177 seats, the Bloc Québecois (BQ—a coalition of a number of parties from the province of Québec) 54, the Reform Party (a new, right-wing party with support derived mainly from the western provinces) 52 and the NDP nine. The Liberals returned to power and Chrétien became Prime Minister.

1 January 1994: The North American Free Trade Agreement (NAFTA), signed in December 1992, came into effect, combining Canada, Mexico and the USA into a free-trade area.

1995: Canada's relations with the European Union (EU) were strained following a dispute over fishing quotas in the north-west Atlantic, particularly following the impounding of a Spanish trawler.

October 1995: A referendum in Québec province narrowly rejected the 'sovereignty' proposed by the ruling PQ.

1996: The Chief of Staff of the Armed Forces relinquished his post, following an inquiry into alleged attempts to suppress information concerning the maltreatment of a Somali citizen by Canadian peacekeeping troops in 1993; further allegations of inappropriate behaviour in Somalia continued to emerge throughout 1996 and 1997.

1997: Canada and the USA failed to agree the annual quota for the fishing of Pacific salmon.

2 June 1997: The Liberals narrowly retained power in a general election, winning 155 of the 301 seats in the enlarged House of Commons. The Reform Party won 60 seats, the BQ 44, the NDP 21 and the PCP 20.

August 1998: The Supreme Court ruled that no province had the right to secede from the federation without prior negotiations and the agreement of the federal government and seven of the 10 provinces. The ruling caused much dissent within Québec.

1 April 1999: Nunavut, Canada's first new territory for more than 100 years, was created as a self-governing homeland for the Inuit people. Formed from part of the Northwest Territories, Nunavut had an initial population of 27,700 inhabitants.

March 2000: Members of the Reform Party voted to join the Canadian Alliance (CA), an organization formed following a convention in September 1998 which aimed to unite the country's major right-wing parties. The PCP subsequently declined to join the CA.

July 2000: Stockwell Day was elected leader of the CA.

October 2000: Chrétien reorganized the Government; shortly afterwards he announced that a general election would be held on 27 November, despite the validity of the current mandate extending until 2002.

27 November 2000: The Liberals won 173 seats in the House of Commons, the CA winning 66, the BQ 38, the NDP 13 and the PCP 12. Chrétien formed a new Liberal Government.

January 2001: The leader of the PQ and Premier of Québec, Lucien Bouchard, announced his resignation, with effect from the election of a successor, claiming that the political climate no longer favoured his preferred route to independence (the BQ having lost numerous seats to the Liberals in the general election held the previous November).

August 2001: Canada became the first country in the developed world to legalize the use of cannabis (marijuana) for medicinal purposes.

Chile

AD 1460: An Inca army invaded the north of Chile and conquered the indigenous Atacameña population.

1491: The Incas established forts in the Central Valley, but failed to subjugate the region. They met fierce opposition from the indigenous Araucanians, in particular from the Mapuche ('the people of the land').

1520: The Portuguese explorer, Ferdinand Magellan, became the first European to visit Chile when he landed on the southern island of Chiloé. Magellan reported that the region was known as Tchili (the Araucanian word for snow).

1535–38: A Spanish expedition, led by Diego de Almagro, failed to discover gold in Chile and returned to Peru.

1540: Pedro de Valdivia led a second expedition into Chile and subsequently founded Santiago.

1553–58: An Araucanian army destroyed a number of Spanish settlements and killed Valdivia. The uprising led to the establishment of the Bío-Bío river as the *de facto* frontier between the Spanish and the Mapuche, an arrangement that lasted for a further three centuries.

18th century: The lack of precious metals and constant conflict with the Mapuche meant Chile remained a relatively unimportant part of the Spanish empire. Santiago remained under the control of the colonial authorities in Lima, Peru.

18 September 1810: The Santiago City Council deposed the colonial Governor of Chile, delegating his powers to a Council of Seven. Spain subsequently dispatched loyal troops from Peru to restore order.

12 February 1817: The royalist army was decisively defeated at Chacabuco, ending Spanish control of northern Chile.

1818: One of the leaders of the independence movement, Bernardo O'Higgins, proclaimed Chile's independence. Royalist forces were not, however, completely expelled from the south of the country until 1826.

1818–23: O'Higgins ruled the country as Director-General of Chile, but opposition to his autocratic style of government eventually forced him to resign.

1823–30: Despite the adoption of a liberal Constitution, there was considerable political instability and no strong government emerged.

1830: Gen. Joaquín Prieto seized control of the Government and, in the following year, was inaugurated as President.

1833: The Government introduced a new Constitution, which granted the majority of power to the executive.

1836: The Peru-Bolivia Confederation was established, with the intention of resisting attempts at expansion by Chile and Argentina, both of which subsequently declared war on the Confederation.

20 January 1839: Chilean forces defeated the Peruvian and Bolivian armies, precipitating the collapse of the Confederation.

1851–61: President Manuel Montt liberalized the Constitution and reduced the privileges of landowners and the Roman Catholic Church.

14 February 1879: Chile invaded Bolivia after the latter had failed to observe the terms of a treaty between the two countries establishing their respective boundaries.

1881: The Chilean army defeated the Mapuche, opening up the south for large-scale European immigration.

1883: The War of the Pacific ended after Chile decisively defeated the Bolivian and Peruvian joint forces. Chile gained the northern territory of Antofagasta from Bolivia, as well as the Peruvian territory of Tarapacá.

August 1891: Forces loyal to the Roman Catholic Church rebelled against the Liberal Government of President José Manuel Balmaceda and defeated a government army in battle.

September 1891: After Valparaíso and Santiago fell to the so-called 'Congressionalist' rebels, Balmaceda committed suicide and Montt became President. The following two decades were marked by serious political instability.

1904: Bolivia and Chile signed the Treaty of Peace and Friendship, by which Bolivia officially ceded its coastal provinces to Chile, thereby losing its access to the Pacific.

August 1906: A massive earthquake virtually destroyed Valparaíso and caused extensive damage to Santiago. An estimated 3,000 people died and a further 100,000 were left homeless.

1920: The Liberal candidate, Arturo Alessandri Palma, was elected President.

September 1924: Alessandri was overthrown in a *coup d'état* organized by members of the armed forces, who subsequently established a military dictatorship.

1925: A second military coup restored Alessandri to power. The armed forces introduced a new Constitution, which reformed the electoral system, reduced the power of Congress and disestablished the Catholic Church.

1926: Emiliano Figueroa was elected President.

1927–31: Gen. Carlos Ibáñez del Campo seized power and established a military dictatorship.

1928: Peru agreed to cede the city of Arica permanently to Chile in exchange for possession of the city of Tacna.

1932: Alessandri was re-elected to the presidency.

1938: Pedro Aguirre Cerda, a member of the Partido Radical, was elected President as the candidate of a coalition of democratic groups.

1939: A devastating earthquake killed about 28,000 people, effectively putting an end to Cerda's ambitious economic recovery plans.

2 April 1942: Cerda died and was replaced as head of state by Juan Antonio Ríos.

27 June 1946: Following a long illness, Ríos died; Alfredo Duhalde became acting President.

3 August 1946: Duhalde resigned as acting President and was replaced by the Minister of the Interior, Vice-Adm. Vicente Merino Bielich.

4 September 1946: The Partido Radical candidate, Gabriel González Videla, won the most votes at the presidential election, and was subsequently appointed President by Congress.

22 August 1947: Congress approved legislation giving the President special powers to solve the country's ongoing economic crisis.

21–22 October 1947: Chile suspended diplomatic relations with several communist countries after the Government claimed that the Partido Comunista had organized a coal-miners' strike. Around 200 prominent communists were also arrested.

16–17 February 1948: Videla inaugurated two army bases on Greenwich Island in order to substantiate Chile's claim to several territories in the South American Antarctic.

4 March 1948: The Chilean and Argentine Governments signed an agreement reaffirming their determination to 'defend their rights in the Antarctic' against any 'aggression' from the United Kingdom.

22 December 1948: Former President Ibáñez was found not guilty of plotting to overthrow the Government and establish a right-wing dictatorship.

6 March 1949: The ruling centrist coalition maintained broadly the same number of seats in Congress at general elections.

16–19 August 1949: The Partido Comunista was banned after the Government claimed it was responsible for serious anti-government rioting in the capital, Santiago.

4 September 1952: Ibáñez was re-elected to the presidency as the candidate of a broad coalition of dissident groups.

1 March 1953: The general election resulted in a 'hung' parliament, with pro-Ibáñez candidates winning exactly one-half of the seats available in the House of Representatives.

3 March 1957: The opposition increased its representation at elections to the House of Representatives; the Partido Radical recorded the largest increase in seats.

August 1958: The ban on the Partido Comunista was lifted.

4 September 1958: Jorge Alessandri, a prominent industrialist and the son of the former President, won the presidential election as the conservative candidate. Alessandri appointed a Cabinet composed largely of technocrats and business-people, with the aim of improving Chile's poor economic performance.

6 April 1959: Congress approved legislation giving special powers to the Government, to allow it to introduce an economic stabilization programme.

1 January 1960: The Government introduced a new currency, the escudo, to replace the peso, which had depreciated significantly in value.

5 March 1960: The Partido Radical performed well at legislative elections and the pro-Government alliance maintained its majority.

May 1960: The south of Chile was devastated by a series of earthquakes, which caused tidal waves and landslides. At least 10,000 people were killed and around 450,000 houses were destroyed.

4 September 1964: The candidate for the centrist Partido Demócrata Cristiano (PDC), Eduardo Frei Montalva, was elected President.

24 November 1964: The Government restored full diplomatic relations with the USSR.

7 March 1965: Following legislative elections, the PDC dramatically increased its representation in Congress.

2 March 1969: The Partido Nacional (an amalgam of the former Liberal and Conservative parties) increased its parliamentary representation, at the expense of the PDC, which lost its overall majority.

4 September 1970: The socialist candidate, Dr Salvador Allende Gossens, won the largest share of the votes at the presidential election and was subsequently appointed President by Congress.

22 October 1970: Right-wing extremists, supported by the US Central Intelligence Agency (CIA), assassinated the Commander-in-Chief of the armed forces, Gen. René Schneider, after Schneider had expressed his belief in military neutrality in political affairs.

3 November 1970: Allende was inaugurated as President and appointed a Government, known as the Unión Popular, which included socialists, Communists, radicals and social democrats.

30 December 1970: President Allende introduced legislation that gave the Government the power to nationalize the country's banking system and coal industry.

4 April 1971: Unión Popular candidates won the majority of the seats contested at local elections.

8 June 1971: The former PDC Minister of the Interior, Edmundo Pérez Zujovic, was assassinated by left-wing extremists. The PDC urged the Government to disarm extremist groups.

11 July 1971: Congress approved legislation to nationalize the copper industry, Chile's most important source of foreign currency. The bill included compensation to be paid to the mines' US owners.

11 October 1971: In a reversal of its previous policy, the Government announced it would not compensate US copper companies and instead asked them to return US $378m. in 'excessive profits'. In response, the US Government threatened to suspend economic assistance to Chile.

10 November–4 December 1971: The Cuban leader, Fidel Castro Ruz, made an official visit to Chile; the visit was fiercely criticized by the centrist and conservative parties.

26 November 1971: Congress rejected a government proposal to merge the two legislative houses into one 'National Congress'.

7 January 1972: Congress impeached the Minister of the Interior, José Tohá González, on the grounds that he had failed to halt the growth of armed paramilitary groups.

26 July 1973: A strike by transport vehicle owners paralyzed the country and prevented supplies of food and petrol from reaching the general public.

13 August 1973: President Allende warned that the country was on the verge of a civil war after more than 22 bombing and shooting incidents were reported in late July and early August.

22 August 1973: The House of Representatives approved a motion of censure against the Government for its handling of the previous month's transport strike.

23 August 1973: Gen. Carlos Prats resigned as Minister of Defence and as Commander-in-Chief of the Army.

11 September 1973: The new Commander-in-Chief of the Army, Gen. Augusto Pinochet Ugarte, launched a military revolt against the Government, during which President Allende died. As President of the Military Junta, Pinochet declared a 'state of internal war'; Congress was dissolved and censorship was introduced.

12 September 1973: Around 7,000 people were arrested by the military. Many of those detained were subsequently tortured and killed.

17 October 1973: The Junta outlawed all political parties and confiscated their assets. The regime subsequently destroyed the country's electoral registers on the basis that they were inaccurate and had been falsified.

30 September 1974: Prats and his wife were killed in Buenos Aires, Argentina, by a car bomb planted by the military regime's secret police, the Dirección de Inteligencia Nacional (DINA).

17 December 1974: Pinochet appointed himself Head of State. He was to hold all executive power, while the Junta would have legislative power.

1 September 1975: The peso again became the official Chilean currency following the repeated devaluation of the escudo.

6 October 1975: The DINA attempted to assassinate former Vice-President Bernardo Leighton and his wife, who were in exile in Rome, Italy.

5 January 1976: The Roman Catholic Church founded the Vicaria de la Solidaridad, which began to document human-rights abuses and to provide legal assistance to victims and their families.

18 February 1976: The US Senate voted to suspend all arms sales to Chile until the Chilean Government improved its human-rights record.

19 February 1976: The UN Commission for Human Rights approved a Resolution condemning the Chilean Government's 'flagrant violations of human rights'.

21 September 1976: A former foreign minister in the Allende Government, Orlando Letelier, and his US assistant, Ronnie Moffat, were killed by a DINA car bomb in the US capitol, Washington, DC.

2 May 1977: The International Court of Justice ruled that several islands in the Beagle Channel, which had been claimed by Argentina, belonged to Chile. Argentina later rejected the decision.

13 August 1977: The Government bowed to US pressure and dissolved the DINA; it was replaced by the Centro de Información Nacional Information Centre (CNI).

5 December 1977: The UN General Assembly condemned the Chilean regime's 'continued and inadmissible violation of human rights'.

4 January 1978: In a 'National Consultation', an estimated 75% of voters reportedly rejected the UN Resolution approved in February 1976 and expressed their support for the Government. The referendum was held in an atmosphere of violent repression.

19 April 1978: The Government introduced an amnesty law to pardon all individuals responsible for crimes between 11 September 1973 and 10 March 1978.

24 July 1978: Gen. Fernando Matthei replaced the Commander-in-Chief of the Air Force, Gen. Gustavo Leigh, as a member of the Junta after Leigh expressed doubts about the Government's policy of repression.

1 August 1978: The US Government requested the extradition of several senior DINA officers on charges relating to the Letelier assassination. The USA introduced military and economic sanctions after Chile's Supreme Court rejected the petition.

8 January 1979: The Chilean and Argentine Governments signed an agreement in Montevideo, Uruguay, which asked the Pope to mediate in the ongoing dispute between the two countries over the possession of islands in the Beagle Channel.

21 March 1980: The President of the Philippines, Ferdinand Marcos, cancelled a state visit by Pinochet, while the President was already on his way to that country. Pinochet subsequently dismissed his Minister of Foreign Affairs, Hernán Cubillos.

12 August 1980: The regime announced that it would hold a referendum on a new constitution, which included an eight-year extension to Pinochet's mandate.

11 September 1980: Some 67% of voters approved the new Constitution in a referendum that was deemed neither free nor fair.

20 October 1980: The PDC President, Andrés Zaldívar, was expelled from Chile after suggesting that the new Constitution was illegal.

20 December 1980: Five Chilean bishops announced the excommunication of torturers, those who ordered torture, and those who could have prevented it but did not do so.

11 March 1981: The new Constitution came into force; Pinochet was sworn in as President.

25 March 1981: Argentina rejected papal recommendations on the Beagle Islands dispute on the grounds that the ruling was more favourable to Chile.

27 February 1982: The leader of the public employees' union confederation, the Agrupación Nacional de Empleados Fiscales (ANEF), Tucapel Jiménez, was assassinated by CNI officers after organizing a general strike in protest at the military regime.

30 August 1983: the Movimiento de Izquierda Revolucionaria (MIR) assassinated the Governor of Santiago, Gen. Carol Urzua Ibáñez.

23 December 1983: The Frente Patriótico Manuel Rodríguez (FPMR) sabotaged central Chile's electricity supply. The Partido Comunista had founded the FPMR to organize armed resistance against the dictatorship.

30 October 1984: A national protest strike was held; the authorities subsequently declared a state of siege and closed down five opposition magazines.

29 November 1984: Following intensive lobbying by Vatican officials, Chile and Argentina signed a treaty of peace and friendship settling the dispute over the ownership of islands in the Beagle Channel.

7 September 1986: The FPMR killed five of President Pinochet's bodyguards during a failed attempt on his life. In response to the assassination attempt, the authorities declared a state of siege.

8 September 1986: Four prominent opponents of the military regime were killed in reprisal for the previous day's assassination attempt.

25 February 1987: Voter registration began for the first time since 1973, in preparation for a referendum on the future of the Pinochet regime, scheduled for the following year.

March 1987: President Pinochet approved a law re-establishing the right to form political parties. Around a dozen parties had registered with the Government by the end of the year.

April 1987: Pope John Paul II visited Chile. Security forces used tear gas against anti-government protesters attending an outdoor mass celebrated by the Pope, who subsequently condemned the regime's record on human rights.

20 October 1987: The Government introduced a law imposing harsh penalties on anyone convicted of belonging to, or reporting the views of, Marxist organizations.

2 February 1988: The 'Comando por el No' political alliance was formed to campaign against the regime in the forthcoming referendum. The PDC leader, Patricio Aylwin, was named as the alliance's spokesman.

30 August 1988: Pinochet was named as the sole presidential candidate in the forthcoming referendum.

22 September 1988: The military regime signed the UN Convention Against Torture, but announced that the accord would not be enacted until March 1990 and would not be applied retrospectively.

5 October 1988: In the national referendum, around 55% of those participating rejected Pinochet's plan to remain in power for a further eight years. Senior military colleagues advised Pinochet to accept the results; it was announced that presidential elections were to be held in December 1989.

14 December 1989: At presidential and legislative elections, Aylwin was elected President with 55.2% of the votes cast, while his centre-left Concertación de los Partidos de la Democracia (as the Comando por el No was now known) won 72 of the 120 seats in the Chamber of Deputies and 22 of the 38 seats in the Senate.

9 January 1990: Aylwin appointed a Cabinet broadly dominated by the PDC and the Partido Socialista (PS).

11 March 1990: At his presidential inauguration, Aylwin announced he would launch an investigation of human-rights abuses under the Pinochet regime.

25 April 1990: The Government established a Truth and Reconciliation Commission under the chairmanship of a prominent lawyer and former Senator, Raúl Rettig.

19 November 1990: A judge investigating a financial scandal involving many prominent businesspeople and senior army officers was killed by a car bomb.

22 January 1991: Congress approved an amendment to the Constitution, which allowed the President to pardon around 250 political prisoners.

4 March 1991: President Aylwin made public the contents of the Rettig Report into human-rights violations under the previous regime. The 1,900-page report documented the deaths of 2,279 people, 957 of whom were classed as 'disappeared' as their bodies had yet to be discovered.

29 March 1991: Pinochet rejected the contents of the Rettig Report, which he described as lacking 'historical or legal validity'.

1 April 1991: The leader of the right-wing Unión Demócrata Independiente (UDI), Senator Jaime Guzmán, was assassinated by the FPMR. The murder led to calls for the Government to increase efforts to disarm extremist groups.

1 June 1991: The FPMR announced it was ending its armed struggle.

7 August 1991: President Aylwin rejected a cross-party proposal that he should extend his term in office by two years, and announced that elections would proceed as planned in 1994.

28 June 1992: The Concertación won 53% of the votes cast at municipal elections, while the right-wing opposition won almost 30%.

September 1992: Four ministers resigned from the Cabinet in order to stand as candidates in forthcoming presidential elections.

10 November 1992: Two former heads of the DINA, Gen. Manuel Contreras and Col Pedro Espinoza, were formally charged with responsibility for the Letelier assassination.

23 May 1993: Senator Eduardo Frei Ruiz-Tagle, son of former President Frei Montalva, won the Concertación's presidential nomination in a primary election in which he defeated the former Minister of Health, Ricardo Lagos Escobar.

12 November 1993: Contreras and Espinoza were sentenced to seven and six years' imprisonment, respectively, for their part in planning the Letelier assassination, thereby becoming the first military officers to be found guilty of human-rights abuses in Chile.

11 December 1993: Frei won an overwhelming victory in the presidential election; at concurrent legislative elections, the Concertación again performed well, although it failed to secure the two-thirds' majority in Congress it needed to reform the military-era Constitution.

14–15 December 1993: Congress approved a motion reducing the presidential term from eight to six years and excluding the possibility of re-election.

21 October 1994: An arbitration court in Rio de Janeiro, Brazil, ended the last remaining border dispute between Chile and Argentina by ruling in favour of Argentine sovereignty over the far-southern Laguna del Desierto.

25 July 1995: The State Defence Council dropped charges of fraud and embezzlement against Pinochet's eldest son, after President Frei warned that the case would have considerable political implications.

11 April 1996: The Senate rejected constitutional reforms proposed by President Frei, which would have forced eight of the nine Senators designated by Pinochet to resign their positions.

17 July 1996: Two right-wing parties, the UDI and the Renovación Nacional (RN), signed an electoral pact.

27 September 1996: President Frei appointed a new Cabinet, which was intended to serve for the remainder of the President's mandate, due to expire in 2000.

27 October 1996: The Concertación again maintained its dominant position at municipal elections.

15 July 1997: The Senate rejected a further government attempt to abolish the eight non-elected seats in the upper chamber.

29 July 1997: The Minister of Housing and Urban Development, Edmundo Hermosilla, resigned after admitting receiving a bribe; he was replaced by Sergio Henríquez.

19 November 1997: The Minister of Defence, Edmundo Pérez Yoma, announced that he would request the resignation of 13 generals as part of the military reorganization, which was to be implemented after Pinochet's retirement as Commander-in-Chief of the Armed Forces.

11 December 1997: At legislative elections, the Concertación retained the same number of seats in the Chamber of Deputies, even though its share of the vote declined. The centre-left parties also won 11 of the 20 seats contested in the Senate.

15 January 1998: Pérez Yoma resigned after failing to persuade Pinochet to bring forward the date of his proposed retirement as Commander-in-Chief.

10 March 1998: Pinochet retired from his military post and took up his place as Senator-for-life, despite widespread protests and several attempts by centre-left politicians to block his incorporation.

9 April 1998: The Chamber of Deputies voted against the impeachment of Pinochet after several PDC deputies voted with the opposition.

4 May 1998: Three mass graves containing the bodies of around 150 political prisoners were discovered near a former detention centre in Pisagua.

16 October 1998: Pinochet was arrested in a London clinic after the British police implemented a warrant against him issued by a Spanish judge investigating human-rights abuses by the military regime.

20–22 October 1998: Around 100 people were arrested following pro- and anti-Pinochet demonstrations in Santiago.

28 October 1998: The High Court in the United Kingdom ruled that Pinochet was entitled to sovereign immunity for actions carried out while he was head of state.

25 November 1998: The British House of Lords ruled that Pinochet was not covered by diplomatic immunity since his actions as head of state had contravened international law. Pinochet remained under house arrest, while the British Secretary of State for the Home Department, Jack Straw, considered whether his extradition to Spain should go ahead. Meanwhile, Pinochet opponents and supporters clashed in Santiago.

27–28 November 1998: The Minister of the Interior, José Miguel Insulza, held talks in London with representatives of the British Government.

17 December 1998: The House of Lords verdict was overturned because one of the Law Lords had not declared his links to the human-rights organization Amnesty International, which was a party to the case. The ruling obliged Straw to suspend the extradition proceedings he had approved against Pinochet.

4 January 1999: The RN and UDI formed a coalition, the Alianza por Chile, and announced that they would present a joint presidential candidate in elections later that year.

24 March 1999: A second British House of Lords panel ruled by a 6–1 majority that Pinochet did not enjoy immunity as a former head of state, but that he should only face charges relating to the period after December 1988, when the United Kingdom ratified the International Convention against Torture.

15 April 1999: The British Home Secretary ruled that extradition proceedings against Pinochet should proceed.

30 May 1999: Ricardo Lagos was elected to be the Concertación's presidential candidate.

21 June 1999: President Frei reshuffled his Cabinet and reappointed Pérez Yoma as Minister of Defence in an attempt to improve relations between the Government and the Armed Forces.

30 June 1999: The US Government declassified thousands of secret files that included details of the CIA's involvement in efforts to destablize the Allende Government.

20 July 1999: A Supreme Court ruling confirmed that abduction was excluded from the terms of the amnesty law allowing prosecutions to go ahead in cases where no body had been found.

14 October 1999: The Chilean Government requested that Pinochet be allowed to return to Chile. The British Government subsequently announced that medical tests would determine if the former ruler were fit to stand trial or should be released.

12 December 1999: In the first round of voting in the presidential election, no candidate won a majority; the two leading candidates, Lagos and the Alianza por Chile nominee, Joaquín Lavín Infante, proceeded to a second ballot.

14 December 1999: The Minister of Justice, Soledad Alvear, resigned from the Cabinet in order to assist with Lagos' presidential campaign.

16 January 2000: Lagos won 51% of the vote at the second round and was elected President. He subsequently announced a Cabinet that was evenly divided between the three main Concertación parties.

3 March 2000: Pinochet returned to Chile after Straw ruled he was too infirm to stand trial.

6 March 2000: The Chilean judge investigating complaints against Pinochet, Juan Guzmán Tapia, asked the Court of Appeal to remove the former President's immunity from prosecution.

23 May 2000: The Court of Appeal ruled that Pinochet should have his immunity from prosecution withdrawn.

13 June 2000: Representatives of the armed forces, the Government and civil and religious institutions agreed to establish a 'Mesa de Diálogo', or round-table discussion, to discuss human-rights issues. Each organization also agreed to search its files for information about the whereabouts of the bodies of the 'disappeared' and produce a report.

21 June 2000: Congress approved government legislation guaranteeing anonymity for members of the military who came forward with information about the 'disappeared'.

8 August 2000: The Supreme Court formally withdrew Pinochet's immunity from prosecution. The ruling also suspended the former dictator's membership of the Senate.

October 2000: Reports in the media suggested that executives at state-owned companies had accepted inflated severance payments at the end of the Frei Government, before being reappointed or transferred to other posts by the Lagos administration. Several executives resigned following the revelations, and Lagos announced that a full inquiry would be held.

29 October 2000: At municipal elections the Alianza por Chile increased its share of the vote to 40%; the grouping's former presidential candidate, Joaquín Lavín, was elected Mayor of Santiago.

8 November 2000: The Supreme Court banned Pinochet and several former DINA officers from leaving the country while the courts considered an extradition request from Argentina relating to the 1974 assassination of Gen. Prats.

9 November 2000: The Supreme Court rejected an extradition request from Italy for Gen. Raúl Iturriaga Neumann, who was accused of organizing the attempted assassination of former Vice-President Leighton in 1975.

18 November 2000: The US Government declassified documents relating to the CIA's involvement in Chile. The files proved that the CIA had financed the PDC and other political groups opposed to Allende.

1 December 2000: Guzmán indicted Pinochet on human-rights charges and placed him under house arrest pending trial.

11 December 2000: The Court of Appeal overturned Guzmán's ruling on the grounds that he had failed to interview Pinochet before indicting him, a necessary step under Chilean law.

26 December 2000: The Supreme Court ruled that Guzmán could not interview Pinochet until the former President had undergone thorough medical examinations.

28 December 2000: The Chilean courts rejected a further extradition request by the Italian Government, which had hoped to try Contreras on charges of involvement in the Leighton assassination attempt.

28 December 2000: President Lagos dismissed the Minister for Housing and Urban Planning, Claudio Orrego; he was replaced by Jaime Ravinet de la Fuente, the former Mayor of Santiago.

6 January 2001: Reports produced by the Mesa de Dialógo provided significant new evidence about the final fate of many of the 'disappeared', although many human-rights campaigners criticized the reports produced by the military and police as 'incomplete and contradictory'.

31 January 2001: Guzmán ruled that medical tests had proved Pinochet was fit to stand trial and he therefore indicted the former President once again, placing him under house arrest.

16 February 2001: The Association of the Families of the Disappeared began legal proceedings against the Commanders-in-Chief of the Armed Forces, accusing them of concealing details about the final fate of the 'disappeared'.

8 March 2001: The Court of Appeal ruled that Pinochet should not face charges of organizing the so-called 'Caravan of Death', a military squad that executed a large number of regime opponents in the weeks following the 1973 *coup d'état*; instead, it ruled that he should be charged with concealing the death squad's activities.

14 March 2001: The Court of Appeal released Pinochet from house arrest on payment of a substantial bail.

21 March 2001: Lawyers loyal to Pinochet began legal proceedings against the surviving members of the Allende Government, accusing them of complicity in left-wing terrorist attacks on military targets.

19 June 2001: The Commander-in-Chief of the Navy, Adm. Jorge Arancibia, provoked widespread outrage by announcing that he was to resign his position in order to stand as a UDI candidate for the Senate.

27 July 2001: There were violent clashes between Mapuche activists and security forces in the southern city of Temuco, as the community's campaign for land restitution intensified.

7 August 2001: The Court of Appeal rejected a further Argentine request for the extradition of Pinochet and several former DINA officers, to face charges of involvement in the Prats assassination.

9 August 2001: The Appeals Court definitively ruled that Pinochet was too infirm to stand trial, effectively ending the human-rights cases pending against him.

13 August 2001: The RN leader, Sebastián Piñera, withdrew his senatorial candidacy rather than compete with Adm. Arancibia, ending an internal dispute within the Alianza por Chile.

28 August 2001: President Lagos asked the Vatican to investigate a speech made by the Chilean Cardinal, Jorge Medina, in which he appeared to urge Roman Catholics to support the UDI in the previous December's legislative elections.

10 September 2001: Chilean and Peruvian government representatives met in Lima, Peru, to discuss co-ordinating the two countries' defence expenditure.

11 September 2001: Relations of the assassinated former Commander-in-Chief, Gen. René Schneider, began legal proceedings in Washington DC against the former US Secretary of State, Henry Kissinger, who was charged with providing CIA support for the attack.

Colombia

c. **1200 BC:** Mesoamerican Indian groups migrated to the area now constituting Colombia from Central America, introducing the cultivation of corn.

AD 1499: At the time of the arrival of the first Spanish forces, Colombia was inhabited by numerous Amerindian tribes, including the Calima, Muiscas, Quimbaya, Sinú and Tayrona peoples.

1500–07: Rodrigo de Bastidas established the first permanent Spanish settlement.

1533: Cartagena was founded by Pedro de Heredia.

1538: Santafé de Bogotá was founded by Gonzalo Jiménez de Quesada. He named the territory that he conquered New Granada, which encompassed present-day Colombia, Ecuador, Panama and Venezuela.

1718: The Viceroyalty of New Granada was established as a separate entity from the Viceroyalty of Peru, which had previously administered the region.

1724–39: The Viceroyalty of New Granada was again incorporated into Peru.

1739: The Viceroyalty of Nueva Granada was restored; Santafé de Bogotá was established as its capital.

1819: Under the leadership of Simón Bolívar, New Granada achieved independence from Spanish rule.

August 1821: Delegates from Venezuela and Colombia met at the border town of Cúcuta to formally sign the Constitution of the Republic of Gran Colombia. The document provided for a presidential system of government, and for a bicameral Congress. Bolívar was confirmed as President, although he left shortly after the signing of the Constitution to conduct military campaigns aimed at the liberation of Bolivia, Ecuador and Peru. The Vice-President, Gen. Francisco de Paula Santander, held executive power in Bolívar's absence.

1822: Upon its liberation from Spanish colonial rule, Ecuador joined the Republic of Gran Colombia.

1827: Bolívar returned to Colombia, although he experienced difficulty in re-establishing his powers.

April 1828: Amid increasing divisions between supporters of Bolívar and Santander, a constitutional convention was convened in order to debate reforms to the 1821 Constitution. No agreement was reached, however, and the divisions continued.

August 1828: Bolívar assumed personal control and attempted to enact a new constitution, naming him President-for-life. This proved highly unpopular, however.

January 1830: During a constitutional convention, Bolívar was forced to resign the presidency; his deputy, José Domingo Caicedo, was installed as his successor.

6 May 1830: Ecuador and Venezuela seceded from Gran Colombia. The remnant, including Panama, was renamed the State of New Granada.

late May 1830: The military intervened, and Gen. Rafael Urdaneta was installed as President.

December 1830: Following Bolívar's death, discontent with the military regime increased and Urdaneta ceded power to Caicedo.

1832: A new Constitution was enacted, restricting the powers of the presidency and providing for a degree of autonomy for the country's regions. Santander assumed the presidency alter in the year.

1837: Santander was succeeded by his deputy, José Ignacio de Márquez.

1840: The country's internal divisions escalated into civil war. Government forces, led by Gen. Pedro Alcántara Herrán, defeated the rebels.

1841: Alcántara Herrán was elected President.

1843: A new, centralist Constitution was enacted, which renamed the country the Republic of New Granada.

1845: Tomás Cipriano de Mosquera was elected President, in succession to Alcántara Herrán. During his tenure, the economy was liberalized.

1848–49: Colombia's two main political parties were formed—the liberal Partido Liberal Colombiano (PL), and the conservative, Partido Conservador Colombiano (PCC).

April 1849: Gen. José Hilario López was elected President; his administration abolished slavery, introduced agrarian reform and formalized the separation of the state and the Roman Catholic Church.

April 1853: Gen. José María Obando y del Campo was elected President.

May 1853: A new Constitution, drafted during López's presidency and formalizing many of his reforms, was enacted.

April 1854: Amid increasing tensions between Conservative and Liberal forces, Gen. José María Melo y Ortiz assumed power, leading a military Government. Gen. Melo dissolved Congress and ruled personally.

December 1854: The PCC deposed Gen. Melo and restored civilian government, under a succession of interim leaders.

April 1857: Mariano Ospina Rodríguez of the PCC was elected to the presidency.

1858: A new Constitution was enacted, renaming the country the Grenadine Confederation.

1860: A PL-led rebellion resulted in severe civil disorder, escalating into civil war.

April 1861: With the conflict preventing the holding of elections, Bartolomé Calvo y Díaz de Lamadrid succeeded to the presidency.

July 1861: Liberal forces led by Mosquera captured Santafé de Bogotá and deposed President Calvo. Mosquera proclaimed himself interim President of the United States of New Granada.

October 1862: Conservative resistance was defeated.

February 1863: A federalist Constitution was enacted by a constitutional convention attended only by Liberals; the country was renamed the United States of Colombia. The high degree of autonomy granted to the units increased the factionalism within the country and resulted in greater internal division. Cipriano was re-elected to the presidency.

1864: Manuel Murillo Toro succeeded Cipriano as President.

April 1866: José Miguel Rojas Garrido became temporary head of state, before the return of Cipriano in late May.

November 1867: Radical Liberal groups deposed Mosquera, who was subsequently exiled. A succession of radical Liberal governments ensued, the presidential term being limited to two years. The governments continued the process of economic liberalization and encouraged the exportation of agricultural commodities, most notably coffee and tobacco.

1876: Marginalized from the political process by the radicalism of the Liberal factions in power, Conservative forces began a rebellion, suppressed by government forces.

April 1880: Rafael Núñez Moledo was elected President. He served until April 1882 and attempted to broaden the Government's support.

February 1884: Following the formation of the Partido Nacional (PN) by Conservatives and moderate Liberals, Núñez was again elected President.

August 1885: A revolt led by radical Liberal elements was defeated. President Núñez declared that the Constitution had expired, and convened a national council to draft a replacement.

1886: A new Constitution was enacted, consolidating the central government and eliminating federalism in order to end rivalries among the numerous political factions. However, this document was considered excessively centralist, in contrast to that of 1863. The PN, under Núñez, became an extreme Conservative group, and suppressed a number of Liberal revolts.

April 1898: Manuel Antonio Sanclemente was elected President. A decline in commodity prices contributed to extreme economic hardship during his tenure.

July 1899–November 1902: A Liberal revolt, subsequently known as the War of the Thousand Days, had initial successes, but was eventually defeated by Conservative forces. During the conflict, in July 1900, Sanclemente was deposed by his Vice-President, José Manuel Marroquín, whose supporters advocated a political solution. Marroquín maintained the conflict, however, and by the time the Liberal forces surrendered, an estimated 100,000 Colombians had died.

1903: A rebellion in the province of Panama was successful; with support from the USA, Panama seceded and became an independent country. Colombia suspended diplomatic relations with the USA in response.

1904: Gen. Rafael Reyes was elected President and attempted to achieve national reconciliation by including members of various groupings in his Government. Reyes abolished Congress, replacing it with an appointed National Assembly.

June 1909: Following presidential attempts to recognize Panamanian independence and restore relations with the USA; opposition to Reyes was concentrated and, in elections to the re-established Congress, a coalition of Liberals and anti-Reyes Conservatives (known as the Unión Republicana) won a majority. Reyes resigned, and was succeeded in the presidency by Carlos Restrepo. Conservative-dominated governments remained in power until 1930.

1919: Relations with the USA were resumed.

1928: Amid the increasing popularity of socialist movements a strike by the workers of the US United Fruit Co. was suppressed by the military.

August 1930: Enrique Alfredo Olaya Herrera, of the PL, was elected to the presidency, following divisions among the Conservatives; Olaya was supported by some moderate Conservatives.

1931: A Peruvian attack on southern Colombia was defeated.

August 1934: Alfonso López Pumarejo, of the PL, was elected President in succession to Olaya. His economic reforms, influenced by contemporary reforms in other American countries, most notably Mexico and the USA, increased the Liberals' popular support.

August 1938: The PL candidate, Eduardo Santos Montejo, was elected to succeed López.

August 1942: López was re-elected, although opposition from leftist elements within the PL and from the Conservatives ensured that he was unable to execute his reformist agenda.

July 1944: President López was detained by military officers during an abortive *coup d'état*; order was restored several days later.

July 1945: López resigned the presidency, and was succeeded by Alberto Lleras Camargo.

August 1946: Leftist elements having gained the ascendancy in the PL, the party's division led to its defeat by the PCC, led by Luis Ospina Pérez.

9 April 1948: The PL candidate for the presidency in 1946, Jorge Eliécer Gaitán, was assassinated, precipitating severe unrest.

1949–58: The most violent of Colombia's civil wars, known as 'La Violencia' and precipitated by the assassination of Gaitán, caused the deaths of about 280,000 people.

June 1953: President Laureano Gómez of the PCC was overthrown in a *coup d'état*; Gen. Gustavo Rojas Pinilla assumed power. Initial popular support for the new regime declined when it became evident that Rojas did not intend to restore constitutional government.

May 1957: A five-man military junta deposed President Rojas.

December 1957: In Colombia's first vote on the basis of universal adult suffrage, an agreement between the PCC and the PL was approved, providing that the presidency was to be held for four-year terms by the PCC and the PL in rotation, with cabinet portfolios and seats in the bicameral Congress to be equally divided between the two parties. This power-sharing agreement, known as the Frente Nacional (FN), was subsequently incorporated into Colombia's Constitution.

May 1958: The FN candidate, Lleras, won the first presidential election under the amended Constitution and the ruling junta subsequently relinquished power. The fighting in 'La Violencia' receded and most of the armed groups either renounced violence or were suppressed.

7 August 1962: Guillermo León Valencia, a Conservative, was inaugurated as President.

1964: The pro-Soviet guerrilla group, the Fuerzas Armadas Revolucionarias de Colombia (FARC), was formed by members of a surviving Violencia militia.

1965: The Ejército de Liberación Nacional (ELN), a pro-Cuban guerrilla group, was established.

7 August 1966: Carlos Lleras Restrepo, a Liberal, was inaugurated as President.

19 April 1970: The FN candidate, Dr Misael Pastrana Borrero, narrowly defeated Gen. Rojas of the Alianza Nacional Popular (ANAPO), to win the presidency. Supporters of ANAPO disputed the results and the armed wing of the party, the Movimiento 19 de Abril (M-19), began to organize guerrilla activity against the Government.

21 April 1974: The bipartisan form of government formally ended. The PL candidate, Dr Alfonso López Michelsen (the son of López Pumarejo—the other two candidates were also the children of former Presidents), won the presidential election, in which support for ANAPO and its candidate, Gen. Rojas Pinilla's daughter, María Eugenia Rojas de Moreno Díaz, was greatly reduced. President López retained Conservative representation in his Government, however.

4 June 1978: Dr Julio César Turbay Ayala of the PL was elected to the presidency.

1982: The problems of urban terrorism and drugs-trafficking became more prevalent and led to a series of successful counter-insurgency operations, aided by an anti-guerrilla group, the Muerte a Secuestradores (MAS), that was itself associated with drugs-trafficking enterprises.

30 May 1982: The PCC candidate, Belisario Betancur Cuartas, won the presidential election with 46.8% of the votes cast. However, a Liberal majority in Congress ensured that an element of consensus would remain a feature of Colombian politics. Betancur altered the country's pro-US orientation, becoming a member of the Non-Aligned Movement.

11 June 1982: The M-19 announced the implementation of a cease-fire.

20 November 1982: A law offering a broad amnesty for guerrillas as part of an internal pacification campaign took effect.

30 April 1984: The assassination of the Minister of Justice, Rodrigo Lara Bonilla, by drugs-traffickers led to a declaration by the Government of a nation-wide state of siege as well as the enforcement of the hitherto unobserved extradition treaty with the USA.

1985–87: A campaign of assassination of Unión Patriótica (UP—a political party formed by FARC in 1985) members by paramilitary 'death squads' resulted in an estimated 450 deaths.

June 1985: The M-19 formally withdrew from the cease-fire agreement.

6 November 1985: A siege by the M-19 at the Palace of Justice, in which 100 people (including 11 judges) were killed, resulted in the indefinite suspension by the Government of negotiations with the M-19.

March 1986: An indefinite cease-fire agreement was concluded between the FARC and the Government.

9 March 1986: The large majority secured by the PL at congressional elections obliged the PCC to form the first formal opposition to a government for 30 years.

25 May 1986: Dr Virgilio Barco Vargas of the PL was elected President with 58% of the votes cast. Barco announced his intention to continue the policy of extraditing drugs-traffickers to the USA; the Medellín drugs cartel began a campaign of violence in an attempt to force the abandonment of extradition.

October 1987: Six guerrilla groups, including the FARC, the M-19 and the Maoist Ejército de Liberación Nacional (ELN), formed a joint front, the Coordinadora Guerrillera Simón Bolívar (CGSB). The cease-fire with the FARC had failed earlier in the year.

late 1987: The Supreme Court ruled that Colombia's extradition treaty with the USA was unconstitutional.

13 January 1989: The Government and the M-19 agreed to resume direct dialogue.

March 1989: The ELN, the EPL and the FARC confirmed their willingness to participate in peace talks with the Government.

August 1989: After a series of assassinations of prominent citizens, allegedly by the drugs cartels of Medellín and Cali, President Barco reactivated Colombia's extradition treaty with the USA. In response to a request by the USA for the extradition of 12 leading drugs-traffickers, the 'Extraditables' declared a 'total war' against anyone opposed to their activities. In the same month, Luis Carlos Galán, who was expected to secure the PL nomination for the presidential contest scheduled to be held in 1990, was assassinated by the Medellín cartel.

October 1989: Following an agreement by the M-19 to demobilize and disarm in exchange for a full pardon, the M-19 was formally constituted as a political party, the Alianza Democrática—M-19 (AD—M-19).

March 1990: Carlos Pizarro of the AD—M-19 became the third presidential candidate to be murdered since August 1989 (the UP candidate, Bernardo Jaramillo, had also been assassinated).

11 March 1990: In congressional elections the PL made substantial gains, winning 72 of the 114 seats in the Senate and a majority in the House of Representatives.

27 May 1990: César Gaviria Trujillo of the PL was elected President with 47.4% of the votes cast. The Medellín cartel subsequently abandoned its campaign of guerrilla attacks, although abductions continued. President Gaviria subsequently offered the cartel immunity from extradition if hostages were released and its members surrendered.

9 December 1990: Elections were held for a 73-seat National Constituent Assembly, convened to devise a new constitution. The PL won 24 seats, the AD—M-19 securing 19 and the PCC only nine.

5 February 1991: The National Constituent Assembly was inaugurated.

19 June 1991: Pablo Escobar, the leader of the Medellín cartel and, consequently, the individual whose extradition had been most fervently sought by the USA, surrendered to the authorities. He was charged with offences including murder and detained in a purpose-built prison outside Medellín.

6 July 1991: The new Constitution was promulgated; extradition was prohibited and emphasis was placed upon provisions to encourage greater political participation and to restrict electoral fraud and corruption.

27 October 1991: In the first congressional elections to be held under the new Constitution, the PL won 58 seats in the 102-seat Senate and 86 seats in the 161-seat House of Representatives.

July 1992: Escobar escaped from prison.

8 November 1992: A 90-day state of emergency was declared following a resurgence of drugs-related violent incidents, many related to the Cali cartel, which had emerged as a powerful force since the partial suppression of the Medellín cartel, although Escobar's escape had precipitated a resurgence in violence.

December 1993: Escobar was killed in an exchange of fire between security forces and fighters loyal to the Medellín cartel.

30 December 1993: A contingent of US troops was dispatched to the Valle del Cauca region ostensibly on humanitarian grounds.

28 February 1994: The US troops were withdrawn after the Council of State ruled that President Gaviria had abused his authority by endorsing their deployment prior to consultation with the Senate.

19 June 1994: In the second round of presidential elections Ernesto Samper Pizano of the PL won 50.4% of the votes cast, defeating Andrés Pastrana Arango of the PCC who obtained 48.6% of the ballot.

17 November 1994: The Government declared its willingness to enter into unconditional dialogue with guerrilla groups.

16 August 1995: Following an escalation in the number of acts of violence perpetrated by the guerrilla groups, the Government declared a 90-day state of emergency.

11 September 1995: Brig.-Gen. Alvaro Velandia Hurtado was removed from active duty—the first action by the Government against senior army personnel accused of human-rights abuses.

28 September 1995: President Samper was forced to protest his own innocence to a congressional committee investigating accusations of improper use of contributions from drugs cartels to fund his presidential election campaign. Samper's former campaign manager was arrested.

2 November 1995: A new 90-day state of emergency was declared after the assassination of a prominent Conservative politician. This was extended in January and April 1996.

1996: The Minister of National Defence, Fernando Botero, President Samper's former election campaign treasurer, Santiago Medina, and a former PL senator, María Izquierdo were all sentenced to terms of imprisonment for their role in the improper use of drugs funds in President Samper's election campaign. The Attorney-General, Orlando Vásquez Velásquez, was convicted of obstructing the course of justice. He was subsequently sentenced to eight years' imprisonment on drugs charges.

12 June 1996: Congress voted to acquit President Samper of charges of having been aware of the part-financing of his election campaign by drugs-traffickers.

11 July 1996: The US Administration accused President Samper of failing to take appropriate measures to deter drugs-trafficking and revoked his visa to travel to the USA.

30–31 August 1996: A major offensive was launched by the FARC and the ELN which resulted in as many as 100 fatalities.

September 1996: The Government mobilized several thousand military reservists to combat the guerrilla groups.

10 September 1996: The Vice-President, Humberto de la Calle, resigned, stating his belief that the Samper Government lacked legitimacy.

May 1997: The Government agreed to the temporary demilitarization of part of the Caquetá department in order to secure the release of 70 members of the armed forces captured by the FARC.

27 September 1997: A total of 652 FARC guerrillas were killed and a further 1,600 were captured in a major military offensive.

November 1997: There was a marked increase in attacks by paramilitary groups; a total of 47 people died in an eight-day period.

8 March 1998: In congressional elections the PL retained a narrow overall majority with 53 of the 102 seats in the Senate and 84 of the 161 seats in the House of Representatives. The PCC secured 27 seats in the Senate and 28 in the House of Representatives.

May 1998: Tens of thousands of Colombians staged a demonstration to protest at the escalation in violence by guerrillas and paramilitary groups.

31 May 1998: Horacio Serpa Uribe, the PL's candidate, was the most popular candidate in the first round of the presidential election with 34.6% of the votes cast; Pastrana, again representing the PCC, received 34.3% of the ballot.

21 June 1998: Pastrana was elected President with 50.4% of the valid votes cast in the second round of voting.

9 July 1998: President-elect Pastrana announced that he had held secret talks with the leader of the FARC, Manuel Marulanda Vélez, and that he had agreed to demilitarize five southern municipalities for a 90-day period.

7 August 1998: Pastrana was inaugurated as President and a new Cabinet was installed.

October 1998: President Pastrana made an official visit to the USA, the first by a Colombian head of state since 1975.

18 October 1998: An ELN attack on a petroleum pipeline in the department of Antioquia resulted in at least 66 civilian deaths.

7 November 1998: The demilitarization of the five southern provinces, with a total area of some 42,000 sq km, began, despite a continuation of FARC attacks on security forces.

mid-January 1999: The peace talks were suspended by FARC, who demanded that action be taken against the paramilitary groups that claimed responsibility for the massacre of more than 130 civilians earlier in the month.

7 January 1999: Preliminary peace talks between the Government and the FARC formally began in San Vicente del Caguán.

February 1999: President Pastrana extended the period of demilitarization by three months. Planned peace negotiations with the ELN did not take place; the ELN attributed this to the Government's refusal to demilitarize four districts in the department of Bolívar.

9 April 1999: Two senior army officers, alleged to have collaborated with the paramilitary groups, were forced to resign.

12 April 1999: The ELN hijacked a domestic flight with 46 people on board.

6 May 1999: Agreement was reached between President Pastrana and Marulanda on a comprehensive agenda for future peace negotiations. The period of demilitarization was again extended.

26 May 1999: The Minister of National Defence, Rodrigo Lloreda, resigned in protest at concessions granted to the FARC.

30 May 1999: The ELN abducted some 140 members of the congregation of a church in Cali.

18 June 1999: President Pastrana withdrew political recognition of the ELN and made the release of hostages a precondition for future peace talks.

17 July 1999: Negotiations between the Government and the FARC were postponed indefinitely after the parties failed to agree on the composition and role of an international verification commission for the demilitarized zone.

19–20 October 1999: Representatives of the Government and the ELN met in Havana (Cuba).

24 October 1999: Negotiations with the FARC were resumed in La Uribe after Pastrana withdrew his demand for international observers.

November 1999: Negotiations with the FARC continued despite a guerrilla offensive against 13 towns.

20 December 1999: The FARC began a 22-day unilateral cease-fire.

29 January 2000: Agreement was reached between the Government and the FARC on the form of future peace negotiations.

20 April 2000: An agreement was reached with the ELN on the establishment of a demilitarized zone in southern Bolívar.

30 April 2000: The FARC sponsored a new political party, the Movimiento Bolivariano para una Nueva Colombia.

May 2000: A cabinet reshuffle was instigated after corruption allegations were made against several ministers; the Government was further reorganized in July.

July 2000: The US Congress approved a contribution of US $1,300m. to a project, known as Plan Colombia (whose total cost was estimated at $7,500m.), intended to strengthen the Colombian state by increasing the efficiency of both the security forces and the judicial system in order to eliminate drugs-production through both eradication and crop-substitution.

30 August 2000: A visit by the US President, William (Bill) Clinton to Colombia was greeted with mass demonstrations protesting against the US role in Colombia.

September 2000: The FARC effectively blockaded the province of Putamayo, Colombia's main coca-growing area, to delay the aerial spraying of coca plantations in that region.

14 November 2000: The FARC withdrew from the peace talks in order to demand greater government action in combating the paramilitary organization the Autodefensas Unidas de Colombia (AUC).

21 November 2000: Colombia recalled its ambassador to Venezuela after a FARC representative was allowed to speak in that country's National Assembly.

January 2001: Plan Colombia officially began.

February 2001: More than 10,000 people in Bolívar participated in a peaceful demonstration against the establishment of a demilitarized zone for the ELN.

8–9 February 2001: President Pastrana and Marulanda agreed to resume formal negotiations. The FARC demilitarized zone was extended for another nine months.

9 March 2001: The ELN suspended peace talks with the Government following an army offensive in Bolívar.

May 2001: The Ministers of National Defence and of Economic Development resigned.

15 May 2001: Some 200 farm workers were abducted by gunmen believed to be affiliated to the AUC, ostensibly in retaliation for the abduction of petroleum workers by the ELN in April.

5–28 June 2001: A series of prisoner exchanges took place between the Government and the FARC.

7 August 2001: Shortly after the resumption of peace talks with the ELN, the Government suspended the negotiations because of a failure to agree on the details for a creation of a demilitarized zone.

12 August 2001: Three Irish nationals were arrested in the FARC-controlled demilitarized zone on suspicion of collaborating with the FARC on training methods. It was alleged that they had links to the Irish Republican Army, which had been conducting a long-standing armed insurgency against British control over Northern Ireland.

24 September 2001: The former culture minister, Consuelo Araujo Noguera, was abducted by the FARC. She was subsequently found murdered.

7 October 2001: The Government extended the FARC's demilitarized zone until 20 January 2002.

Costa Rica

AD 1502: The Italian-born explorer in the service of the Spanish Crown, Cristoforo Colombo (Christopher Columbus) arrived in the country and named it Costa Rica (rich coast), after the gold and treasure that was rumoured to be found there.

1522: The Spanish army moved south and colonized the Meseta Central. Large numbers of the indigenous population succumbed to diseases such as smallpox and influenza, introduced by the Europeans.

1563: The city of Cartago was founded.

1570: Costa Rica was administered by the Captaincy-General of Guatemala.

1737: San José, the present-day capital of Costa Rica, was founded.

1808: Coffee began to be cultivated.

1821–23: The colony declared its independence from Spain and became incorporated into the newly-formed Mexican Empire.

1823: Costa Rica joined the United Provinces of Central America.

1838: Following the collapse of the Central American federation, the country became an independent republic.

1849–59: Under the dictatorship of J. Rafael Mora, coffee became the country's major export.

1870–82: Following a bloodless *coup d'état*, a liberal government was formed under the leadership of Tomás Guardia.

1889: The first democratic elections were held under President Bernardo Soto.

1890: President Alfredo González was deposed by Federico Tinoco, who assumed the presidency under a new Constitution. The USA refused to recognize Tinoco's revolutionary Government, however. Following a counter-revolution, Julio Acosta won a presidential election, and relations between the two countries were restored.

March 1948: Presidential elections were characterized by violent protests, including a 15-day general strike. When the opposition candidate, Otilio Ulate Blanco, was declared victorious, the Government's candidate, Rafael Angel Calderón Fournier, contested the result, which was eventually nullified. José Figueres Ferrer then led a revolt in support of Ulate. Santos León Ferrera was appointed interim President.

May 1948: The Constitution was abolished and the socialist Figueres and his junta assumed control of the Government.

January 1949: A new Constitution, which abolished the army, was formulated.

November 1949: The junta relinquished power and Otilio Ulate was elected President.

1952: Figueres withdrew his support for Ulate and formed the Partido de Liberación Nacional (PLN).

1953: Figueres was elected President.

1958: The conservative candidate, Mario Echandi Jiménez, was victorious at presidential elections.

1962: Jiménez's successor, Francisco Orlich Bolmarich, was appointed President following elections.

1966: The conservative candidate José Joaquín Trejos Fernández, became President.

1970: Figueres was re-elected as President.

1974: Figueres' successor, Daniel Oduber Quirós, took office as President.

1978: Rodrigo Carazo Odio of the conservative Partido Unidad Opositora (PUO) coalition (later the Coalición Unidad) was elected President.

1981: Odio was criticized for his alleged involvement in the illegal trafficking of arms between Cuba and El Salvador.

7 February 1982: Presidential elections resulted in victory for the PLN candidate, Luis Alberto Monge Alvarez. In concurrent legislative elections, the ruling conservative Coalición Unidad came second, obtaining 34% of votes cast; the PLN won a majority, securing 33 of the 57 seats in the Legislative Assembly (Asamblea Legislativa).

May 1982: Monge's administration took office and was confronted with domestic economic crises and external conflicts in the neighbouring regions, particularly Nicaragua.

1983: The President officially declared the country's neutrality with regard to the conflict in Nicaragua.

December 1984: Relations between Costa Rica and Nicaragua deteriorated when the Government accused Nicaragua of violating international law, following an incident involving a Nicaraguan refugee at the Costa Rican embassy in Managua. Diplomatic relations were subsequently reduced to a minimum.

February 1986: The PLN candidate, Oscar Arias Sánchez, won a presidential election. The PLN was also successful at legislative elections, held concurrently. Later that month, diplomatic relations with Nicaragua were restored.

May 1986: President Arias announced his intention to pursue a policy of neutrality; none the less he also recognized the necessity of maintaining good relations with the USA, to ensure an influx of foreign aid. However, the US administration subsequently reduced aid to Costa Rica, a decision that was widely regarded as an attempt by the USA to coerce the Government into supporting the Nicaraguan counter-revolutionary group, the Contras.

August 1987: Costa Rica, El Salvador, Guatemala, Honduras and Nicaragua signed a peace accord proposed by President Arias (who was awarded the Nobel Peace Prize later that year).

January 1988: President Arias organized the first meeting between Nicaraguan government officials and Contra leaders.

August 1988: Internal unrest and strikes occurred as a result of the Government's 'Agriculture for Change' policy, that promoted the cultivation of cash crops.

1989: Unrest increased as trade unions, professional bodies and civic groups united to demonstrate against the Government's policies of structural adjustment. In September the Minister of Finance, Fernando Naranjo, resigned after his efforts to impose austerity measures were undermined by a budgetary deficit. Also in that month, a report by the Asamblea into the extent of drugs-trafficking implicated many political and business figures, including the former PLN President, Daniel Oduber Quirós, of involvement in illegal activities; Quirós was later obliged to resign.

January 1990: Diplomatic relations with Panama were normalized following the overthrow by US forces of Gen. Manuel Noriega.

February 1990: Calderón Fournier, of the Partido Unidad Social Cristiana (PUSC), was victorious at presidential elections. The PUSC also secured a majority in the Asamblea.

1993: Social and political tensions arose owing to the Government's privatization programme and reductions in expenditure in education and in the public health sector. Measures including the privatization of the telecommunications and health sectors, and of the state petroleum company, Refinadora Costarricense de Petróleo (Recope), were opposed by the PNL.

6 February 1994: José Maria Figueres Olsen, the son of José Figueres Ferrer, was elected as President and pledged to eradicate poverty.

March 1994: The Government signed a free-trade agreement with Mexico, which granted approximately 86% of Costa Rican exports duty-free access to the Mexican market.

April 1995: The Government made an agreement with the opposition PUSC to ensure the implementation of tax increases and austerity measures. Relations with Nicaragua were again strained following allegations that a group of illegal Nicaraguan immigrants had been violently expelled from Costa Rica.

1996: Social unrest led to the suspension of plans to privatize the energy and telecommunications sectors. Figueres' popularity decreased dramatically as living standards declined, taxes increased, and social expenditure was significantly reduced.

1997: Despite an improvement in economic conditions, President Figueres' popularity did not recover and deep divisions appeared within the PLN.

1 February 1998: Presidential elections resulted in victory for the PUSC candidate, Miguel Angel Rodríguez Echevarría, who received 46.9% of votes cast, thereby narrowly defeating the PLN candidate, José Miguel Corrales Bolaños. The PUSC failed to obtain an outright majority in the legislature, securing 27 of a total 57 seats in the Asamblea; the PLN won 23, the coalition Fuerza Democrática (FD) three, the Partido Movimiento Libertario (PML) two, and the Partido Integración Nacional (PIN) and the Partido Acción Laborista Agrícola (PALA) both secured one seat.

July 1998: Relations with Nicaragua further deteriorated when Nicaragua prohibited Costa Rican civil guards from carrying arms while navigating the San Juan river, which forms the border between the two countries.

February 1999: A six-month amnesty was declared for all illegal Nicaraguan immigrants who had entered Costa Rica prior to November 1998.

March 2000: Arias campaigned for a constitutional amendment to enable him to stand in the 2002 presidential elections. However the amendment did not receive the support of the PUSC, and Arias subsequently lodged an appeal with the Constitutional Court.

April 2000: Popular protests forced the Government to withdraw legislation opening the energy and telecommunications industries to private investment. The Supreme Court rejected the proposed law on procedural grounds.

June 2000: The Government's dispute with the Nicaraguan Government regarding the San Juan river appeared to have been resolved when both countries agreed a procedure that would allow armed Costa Rican police officers to patrol the river.

September 2000: The Constitutional Court ruled against Arias, ending his hopes of standing for the 2002 presidential elections.

May 2001: Renewed disputes with Nicaragua arose following the implementation of a US $25 charge for Costa Ricans using the San Juan river.

June 2001: A constitutional reform commission was appointed by President Rodríguez to determine the changes required to convert the current presidential system of government into a semi-parliamentary one.

July 2001: A dispute arose with Nicaragua over a wall constructed on the frontier between the two countries, used for the control of border traffic.

Cuba

***c*. 1000 BC:** The Guanahacabibe people settled the island.

***c*. AD 900:** A group of Arawak Indians, the Taínos, invaded the island, establishing settlements.

27 October 1492: The Italian navigator in the service of the Spanish Crown, Cristoforo Colombo (Christopher Columbus), discovered the island and claimed it for the Spanish crown.

1511: Diego de Velázquez led a fleet of Spanish ships from the colony of Hispaniola, which landed at what is now known as Guantánamo. The Spanish subsequently defeated the native chief, Hatuey. At the same time, Pánfilo de Narváez led an expedition to the southern coast; converting the Taíno to Roman Catholicism and obliging them to recognize the Spanish king as their monarch.

1542: The indigenous population had declined dramatically by this time, owing to harsh working conditions, malnutrition and their susceptibility to diseases introduced by the Europeans.

1555: A French privateer, Jacques de Sores, attacked the island's capital, Havana. The eastern coast suffered from incursions by French and British pirates.

1717: Tobacco growers rebelled against the Spanish authorities' monopoly on the tobacco trade.

1762: Havana was attacked by a British fleet that then occupied the island for 10 months; however, it was subsequently returned in exchange for Florida. The Spanish Government relaxed restrictions that had previously prevented the colony trading with any territory other than Spain.

1848: Growing concern amongst plantation owners for the preservation of the institution of slavery caused a group of 'annexationists' to request that Cuba become part of the USA, where slavery remained legal in the southern states. The Spanish Government, however, rejected the US President's offer of US $100m. for the island.

1854: A further offer of US $130m. by the US Government for Cuba was also refused by Spain.

10 October 1868: A landowner, Carlos Manuel de Céspedes, freed his slaves and led a rebellion by activists demanding social reform, effective representation, freedom of speech and association, and racial equality.

1869: The rebels convened a Constituent Assembly at Guaímaro, at which they announced their intention to form a revolutionary government and achieve independence.

1878: The Ten-Year War (or Yara Revolution) ended following the signing of the Zanjón Pact, which freed rebel slaves and extended political rights.

26 August 1879: Calls for the full abolition of slavery and complete independence developed into what became known as the Little War (La Guerra Chiquita), that lasted for two years.

1886: Slavery was abolished.

April 1892: The Partido Revolucionario Cubano (PRC) was formally established. It was led by the writer, José Martí, who solicited support among separatists in the USA, as well as on the island itself.

1895: The PRC began a war of independence in the east of the island. Spain responded by sending troops, led by Gen. Valeriano Weyler, who later ordered the construction of detention camps, in which thousands of rebel supporters died.

1896: In response to public alarm at the atrocities committed by Weyler, the President of the USA, William McKinley, ordered the Spanish Government either to enact reforms, win the war, or accept the intervention of the USA.

April 1898: Following the sinking of a US battleship in Havana harbour, the US Government declared war on Spain. US navy vessels blockaded Havana and its troops laid siege to Santiago de Cuba.

10 December 1898: The Treaty of Paris was concluded between the Spanish and US Governments, formally ending the Spanish–American War. The terms of the treaty bestowed sovereignty over the colony on the US Government.

1 January 1899: The US Government formally began its rule of Cuba. The patriot army having been disbanded, their original demands for racial and social equality were ignored.

1901: A new Constitution was drafted, providing for the establishment of a Congress, comprising two houses. Local, provincial and national leaders were to be democratically elected. The US authorities also insisted on a series of appendages to the Constitution, known as the Platt Amendment, prohibiting Cuba from making treaties and alliances with other countries, and allowing the USA to maintain military bases on the island. The leader of the PRC, Tomás Estrada Palma, was elected President.

20 May 1902: Cuba was granted independence.

1905: A rebellion occurred, organized by liberals disputing the re-election of Estrada Palma to the presidency.

August 1906: Estrada Palma and his Cabinet resigned; US marines, led by Charles E. Magoon, arrived to commence the task of restoring law and order.

1908: US troops withdrew.

1917: The election of the Partido Conservador candidate, Mario García Menocal, to the presidency provoked a rebellion by the Partido Liberal, that accused its opponents of electoral fraud.

1920: The Liberal Party candidate, Alfredo Zayas, was elected President.

November 1924: Gerardo Machado of the Liberal Party was elected President. He revised the Platt Amendment, imposed taxes on imported products and controlled sugar production.

1928: Machado was re-elected unopposed for an extended six-year term. This provoked opposition from radical student groups, whcih began anti-Government campaigns throughout the country.

1930: Machado decreed spontaneous demonstrations to be illegal. A special police force was established to arrest, torture and imprison anti-Government protesters.

1932: Machado suspended the Constitution.

September 1933: A general strike led the armed forces to lose support in Machado, compelling him to resign from the presidency. The US diplomat, Sumner Welles, who had been mediating between the Government and the opposition, appointed Carlos Manuel de Céspedes (the younger) as President; however Sgt (later Gen.) Fulgencio Batista y Zaldívar led a *coup d'état*, overthrowing Céspedes, and installed Ramón Grau San Martín as President.

January 1934: Grau resigned following the refusal by the USA to recognize his administration, and Batista assumed control of the administration himself.

1940: A new Constitution emphasizing democracy and social justice came into effect and Batista was officially elected President.

1944: Batista retired and, as leader of the Partido Auténtico, Grau returned to the position of President.

1948: Carlos Prío Socarrás was elected President.

10 March 1952: Batista seized control in a *coup d'état*, took the title of Chief of State and revoked the 1940 Constitution.

4 April 1952: Batista installed himself as Provisional President, pending presidential elections.

26 July 1953: An attempt by a group of revolutionaries, led by Fidel Castro Ruz, to seize control of military barracks in Santiago de Cuba, was brutally suppressed. Castro was subsequently imprisoned.

1954: Batista was re-elected President; he subsequently released Castro into exile.

December 1956: Castro and his followers unsuccessfully attempted to invade Cuba. Castro and an Argentine-born revolutionary, Ernesto 'Che' Guevara , subsequently formed an anti-Government guerrilla force.

13 March 1957: Members of a group of intellectuals, known as the Directorio Universitario, led by José Antonio Echeverría, attempted to assassinate Batista.

1 January 1959: Batista fled the country and Castro's forces occupied Havana. The Constitution was subsequently suspended and replaced by a Fundamental Law, and a Council of Ministers was appointed.

March 1959: Fidel Castro assumed the office of Prime Minister, with his brother, Raúl Castro, as his deputy.

October 1960: Castro's administration expropriated all US business interests in the country.

January 1961: The Government terminated diplomatic relations with the USA.

April 1961: A US-sponsored group of anti-Castro Cuban émigrés landed at the Bahía de Cochinos (Bay of Pigs) in the South; the invasion was repelled by Castro's troops.

December 1961: Castro proclaimed Cuba to be a Communist state. The Organizaciones Revolucionarias Integradas (ORI, later renamed the Partido Unido de la Revolución Socialista Cubana—PURSC) was declared to be the state's sole legal political party.

January 1962: The US Government imposed an economic blockade on the country, and Cuba was excluded from the Organization of American States (OAS).

October 1962: The discovery of Soviet nuclear missiles on Cuba provoked a grave international crisis. Following a naval blockade by US forces, the missiles were withdrawn.

1964: The OAS imposed diplomatic and commercial sanctions against Castro's Government.

October 1965: The PURSC was replaced by the Partido Comunista de Cuba (PCC).

July 1972: The Government became a member of the Council for Mutual Economic Assistance (CMEA) and received preferential trade terms and technical advice from the USSR and other Eastern European countries.

February 1976: A new Constitution was promulgated, following its approval in a popular referendum.

December 1976: Following elections, the Asamblea Nacional del Poder Popular (National Assembly of People's Power) was inaugurated as the supreme legislative organ of state. The delegates in turn elected a Council of State, with Fidel Castro as President.

November 1981: The Government held high-level talks with the US Government; despite this, US hostility increased. The following year economic sanctions were strengthened, a major air link was terminated and tourism and investment by US nationals was prohibited.

December 1981: Castro was re-elected President of the Council of State.

December 1984: Following further negotiations with the US Government, an agreement was reached on the resumption of immigration and the repatriation of 2,746 Cuban 'undesirables' who had emigrated to the USA in 1980.

February 1985: The Government suspended the programme of immigration and repatriation, in response to the establishment of a radio station in Florida (USA), which broadcast Western-style programmes to the island. All visits to Cuba by US residents of Cuban origin were subsequently banned.

February 1986: At the third Congress of the PCC, one-third of the 146 members were replaced.

September 1986: As the result of mediation by the Roman Catholic Church, political prisoners and their families were permitted to emigrate to the USA.

October 1987: The Government's decision to reinstate the 1984 accord on immigration and repatriation provoked rioting by Cuban exiles detained in US prisons; the protests were quelled only when the inmates were assured that they would not be repatriated.

May 1988: Government troops launched a successful offensive against South African forces present in Angola. Castro had been assisting the rebel Angolan forces since 1976.

April 1989: Following the declaration of peace in Angola, Cuban forces began to withdraw.

April 1989: The President of the USSR, Mikhail Gorbachev, met Castro for talks regarding the reduction of Soviet aid and Central American issues. Castro subsequently denounced Gorbachev's policies of *perestroika* (restructuring) and *glasnost* (openness) as causing crisis in socialism.

June 1989: Several senior officers were found to have been involved in smuggling operations in Angola and to have assisted the Colombian Medellín drugs-trafficking cartel. Gen. Arnaldo Ochoa Sánchez and three others were convicted of treason and executed.

9 September 1989: Cuba announced the full withdrawal of its troops from Ethiopia by September 1990.

October 1989: The country was elected to the UN Security Council.

25 January 1990: Cuba suspended troop withdrawals from Angola, following the killing of four Cuban soldiers by guerrillas; withdrawals were resumed in February.

November 1990: Owing to the adverse effects of the US economic embargo, rationing was extended to all products.

25 May 1991: The final contingent of Cuban troops was withdrawn from Angola.

September 1991: The Soviet Government announced that it would withdraw the majority of its military personnel from Cuba. Castro's subsequent demands for the withdrawal of US troops from Guantánamo were denied.

July 1992: The Asamblea Nacional approved amendments to the Constitution, which empowered Castro to declare a state of emergency and assume full control of the armed forces, as the head of a National Defence Council.

October 1992: The Government imposed the Cuban Democracy Act ('Torricelli Law'), which made it illegal for foreign subsidiaries of US companies to trade with Cuba. This was condemned by the international community and by the UN General Assembly.

24 February 1993: Elections to the Asamblea Nacional and 14 provincial assemblies took place and, for the first time, were conducted by secret ballot. However, only candidates nominated by the PCC were permitted to stand for election. All deputies to the Asamblea were re-elected. The following month, Fidel Castro and Raúl Castro were re-elected as President and First Vice-President respectively.

July 1993: Castro announced that the prohibition of the possession of foreign currency by Cubans was to be lifted, and that restrictions on Cuban exiles travelling to Cuba were also to be relaxed.

September 1993: The Government authorized limited individual private enterprise in a range of occupations. Agricultural reforms were also introduced to decentralize the administration of state farms to form 'Units of Basic Co-operative Production', that were to be managed by the workers themselves.

April 1994: In a Government reorganization, several new ministries were created, including those of economy and planning, finance and prices, and foreign investment and economic co-operation.

August 1994: A deterioration in economic conditions prompted a mass migration to the USA; in response to the crisis, the President of the USA, William (Bill) Clinton, suspended the automatic refugee status conferred on Cubans by the 1966 Cuban Adjustment Act.

September 1994: The Government and the USA held talks regarding the refugee crisis. The US Government pledged to grant visas allowing for the migration of a minimum of 20,000 Cuban refugees annually, in exchange for the reintroduction of travel restrictions by Cuba.

May 1995: The USA officially revoked the automatic refugee status enjoyed by Cubans; it was also announced that all Cuban refugees intercepted at sea by US naval forces would be forcibly repatriated.

September 1995: The US Congress approved legislation, known as the Helms-Burton, which sought to impose sanctions on countries trading with or investing in Cuba and threatened to reduce US aid to the country's trading partners. In response to international condemnation the Senate eventually passed a modified version of the original bill.

February 1996: Cuban air force fighters shot down two US light aircraft piloted by members of Brothers to the Rescue, a Cuban-American exile group, killing all four crew members. The USA imposed further economic sanctions and indefinitely suspended charter flights to Cuba.

12 March 1996: The Cuban Liberty and Democratic Solidarity Act, as the Helms-Burton Bill was entitled, was enacted. Title III, permitting US citizens to prosecute any foreign corporation or investor with business dealings involving property expropriated by Castro's regime, was condemned by the international community.

June 1996: Canada, Mexico and the European Union (EU) announced that they had imposed legislation to protect their businesses from the implications of Title III. The following month, the US Government announced that a six-month moratorium had been imposed on the legislation.

November 1996: The UN General Assembly voted in favour of repealing the USA's sanctions on Cuba; the World Trade Organization (WTO) announced the formation of a disputes panel to rule on the legality of the Cuban Liberty and Democratic Solidarity Act.

11 January 1998: Elections to an enlarged Asamblea Nacional and to the 14 provincial assemblies took place, and all candidates were elected.

22–25 January 1998: The Government received a visit from the Pope, who condemned the US embargo and recommended that the country should be reintegrated into the international community. Castro subsequently released a number of political prisoners.

24 February 1998: Fidel Castro was re-elected as President and Raúl Castro as Vice-President.

March 1998: The US Government permitted the shipment of food and medicines to the country and announced the lifting of the ban on direct flights.

September 1998: The Asamblea Nacional condemned the US economic sanctions as an act of 'genocide'.

October 1998: At the annual vote by the UN General Assembly on the Resolution condemning US economic sanctions, all but two members (USA and Israel) voted against the sanctions.

November 1999: A five-year old Cuban boy, Elián Gonzalez, was rescued from a sinking vessel of refugees bound for the USA. The ensuing custody battle between the boy's relatives in Miami and his father in Cuba, further strained Cuba's relations with the USA.

April 2000: The annual US-sponsored vote of censure against the Cuban Government for alleged human rights abuses, held at a meeting of the UN Commission on Human Rights, was defeated for the first time since 1991.

April 2000: Despite a ruling by the US Department of Justice granting custody to Elián's father, his Miami relatives refused to allow the boy to be repatriated. Federal Bureau of Investigation agents subsequently forcibly removed him from their control in preparation for his return to Cuba in June, where the case had provoked public demonstrations against the USA's 'abduction' of the child.

15 December 2000: Following the imposition by the Government of a 10% tax on all telephone calls to the USA, the latter suspended telephone links.

12 January 2001: A member of the Czech legislative, Ivan Pilip, and a Czech political activist, Jan Bubenik, were arrested following allegations that they had held meetings with anti-Government dissidents. They were released the following month.

4 May 2001: The Ministry for Auditing and Control was established to combat government corruption.

14 July 2001: The US President, George W. Bush, announced that the US trade embargo would be further reinforced. On 16 July he announced that Title III was to remain suspended.

8 November 2001: Following the destruction of thousands of homes and the deaths of five people in hurricane Michelle, the US Government offered to send humanitarian aid, on the condition that it did not assist the Castro Government. Castro had previously rejected such aid from the USA.

Dominica

***c.* 3100 BC:** The Ortoroid people from South America gradually settled in the Caribbean area.

***c.* 400 BC:** The Ortoroid people disappeared.

AD 400: The Arawak Igneri people settled in what is now Dominica.

1400: The Kalinago tribe, known as the Caribs, settled in the Caribbean area, ousting the Igneri from the island, which they named Waitikubuli ('tall is her body').

3 November 1493: The Italian navigator under the service of the Spanish Crown, Cristoforo Colombo (Christopher Columbus), sighted the island on a Sunday, and thus named it Dominica. The Caribs resisted attempts by the Spanish to colonize the island.

early 1600s: The British and the French arrived in Dominica, but agreed that the island should remain neutral.

1627: The English monarch, King Charles I, placed the island in the charge of the Earl of Carlisle, however, attempts at colonization were resisted by the Caribs.

1635: France laid claim to Dominica, but this met with strong resistance by the Caribs.

1660: England and France agreed to abandon their claims to Dominica.

1761: British forces attacked French settlements in Dominica, claiming the territory.

1763: Following the Treaty of Paris which ended the Seven Years' War, Dominica was officially ceded to the British and a Legislative Assembly was established.

1778: The French attacked Dominica, taking control of the island.

1782: The naval Battle of the Saintes returned Dominca to British rule. Dominica subsequently formed part of the Leeward Islands federation.

1834: The United Kingdom abolished slavery throughout its Empire.

1865: The United Kingdom replaced the elected Legislative Assembly with one consisting of elected and appointed members in equal numbers.

1896: Crown Colony government was re-established in Dominica by the United Kingdom.

1939: Dominica became part of the Windward Islands federation.

2 August 1956: The British Parliament adopted a bill establishing the British Caribbean Federation, which was to have a joint Government, Legislature and Supreme Court.

1958: The West Indies Federation was created; this included almost all of the United Kingdom's Caribbean colonies and had the function of assisting these in moving towards independent statehood.

6 October 1959: The British Colonial Office announced the abolition of the post of Governor of the Windward Islands as part of the reform of the islands' constitutions. The reforms were designed to reduce the administrative differences between the members of the West Indies Federation.

17 January 1961: The ruling Dominica United People's Party was defeated by the Dominica Labour Party (DLP) at a general election, and the DLP's leader, Edward Le Blanc, became Chief Minister.

27 February–5 March 1962: Representatives of Barbados and the seven Leeward and Windward Island colonies met in Barbados and requested that the British Government approve a new federation, known as the Little Eight, in the event of the collapse of the West Indies Federation.

31 May 1962: The British Government formally dissolved the West Indies Federation.

18 April–4 May 1966: A Constitutional Conference was held in London to discuss the future of Dominica, Grenada, St Christopher-Nevis-Anguilla, St Lucia and St Vincent, following the collapse of the proposed Little Seven federation (as it had become known after Grenada's withdrawal).

1 March 1967: Dominica was granted the status of an Associated State and became self-governing, although the United Kingdom retained responsibility for defence and foreign relations. The Legislative Council was replaced by the House of Assembly, the Administrator became Governor and the Chief Minister was restyled Premier.

26 October 1970: Edward Leblanc was re-elected Premier after his faction of the Labour Party, the Leblanc Labour Party, won eight of the 14 seats in the House of Assembly.

25 July 1972: The Governments of Guyana, Dominica, Grenada, St Christopher-Nevis-Anguilla, St Lucia and St Vincent signed the Declaration of Grenada, which created the framework for an Eastern Caribbean political union.

27 July 1974: Leblanc resigned as Premier and was replaced by the then Deputy Premier and Minister of Finance, Patrick John.

24 March 1975: The Dominica Labour Party (DLP) won 16 of the 21 elective seats in the House of Assembly and John was reappointed Premier.

3 November 1978: Dominica became an independent self-governing republic under the name of the Commonwealth of Dominica. John became the island's first Prime Minister, and the then Speaker of the House of Assembly, Frederick Degazon, was eventually elected President.

29 May 1979: Two people were killed by the Defence Force at a demonstration against government attempts to introduce legislation restricting the freedom of trade unions and the press. Trade unions subsequently organized a general strike and a pressure group, the Committee for National Salvation (CNS), to call for John's resignation.

June 1979: President Degazon fled Dominica after John rejected his advice that he resign. All of the members of John's Cabinet resigned, and John eventually agreed to relinquish power. The candidate proposed by the CNS, Oliver Seraphin, was elected Prime Minister and an interim Government was formed.

February 1980: President Degazon resigned and Aurelius Marie was appointed as the new President.

13 February 1980: The Government declared a state of emergency following a kidnapping carried out by a radical organization known as the Dreads, in retaliation for a police crackdown on marijuana.

21 July 1980: The conservative Dominica Freedom Party (DFP) achieved a significant victory at legislative elections, obtaining 17 of the 21 elective seats in the House of Assembly. The DFP's leader, Mary Eugenia Charles, was appointed Prime Minister.

7 March 1981: The Government announced that the Dominica Defence Force would be disbanded after several of its officers were implicated in a plot to overthrow the Government.

October 1981: A number of people, including John, were committed for trial on charges of plotting to overthrow Charles' Government.

19 December 1981: The Government declared a state of emergency after attacks were launched on the island's prison and police stations. The plot, which was defeated by the security services, was intended to free John and several other prisoners.

May 1982: John and his fellow defendants were acquitted of seeking to overthrow the Government, owing to a lack of evidence.

10 January 1983: The Government announced that it was to extend the state of emergency for a third year.

19 December 1983: Marie completed his five-year term as President and Clarence Seignoret was elected to replace him.

17 January 1985: The DLP and the United Dominica Labour Party (UDLP) merged to form the Labour Party of Dominica (LPD) under the leadership of the UDLP leader, Michael Douglas.

1 July 1985: The ruling DFP was re-elected at legislative elections, securing 59% of the votes cast and 15 of the 21 seats in the House of Assembly; the opposition LPD obtained five seats. Charles was reappointed Prime Minister. The LPD began an 18-month boycott of the House, in protest at the Government's decision to curtail live broadcasts of parliamentary proceedings.

23 October 1985: John was sentenced to 12 years' imprisonment for attempting to overthrow the Charles Government. Several other defendants also received prison terms.

8 August 1986: The former Commander of the Dominica Defence Force, Maj. Frederick Newton, was hanged after being found guilty of the murder of a police officer during the December 1981 attempts to free John.

20 December 1988: Seignoret was re-elected as President.

28 May 1990: At a general election, Charles narrowly secured a third term in office after the DFP won 11 of the 21 elective seats in the House of Assembly; the UWP became the largest opposition party, securing six seats, and the LPD obtained four seats.

18 April 1992: Charles narrowly survived a vote of 'no confidence' brought by opposition members who accused the Government of discriminating against its political opponents.

25 October 1993: The then Speaker of the House of Assembly, Crispin Sorhaindo, replaced Seignoret as President, following the latter's retirement from office.

26 April 1994: Charles introduced a state of emergency and a curfew in the capital, Roseau, following a series of anti-Government demonstrations.

12 June 1995: The UWP narrowly defeated the ruling DFP in legislative elections. The UWP leader, Edison James, was appointed Prime Minister in place of Charles, who announced her retirement from politics.

2 October 1998: The House of Assembly elected Vernon Shaw, a former Cabinet Secretary, as President.

31 January 2000: The DLP won a narrow victory at legislative elections, obtaining 10 seats and 42.9% of votes cast; the UWP secured nine seats and 43.4% of votes and the DFP won two seats. The DLP leader, Roosevelt (Rosie) Douglas, formed a coalition Government with the DFP and was appointed Prime Minister.

26–27 June 2000: The Organization for Economic Co-operation and Development (OECD) and the Financial Action Task Force (FATF, part of the OECD) criticized Dominica's financial system. The OECD requested that the country reform its rules by 31 December 2005 or face sanctions.

4 July 2000: The Minister of Agriculture, Planning and the Environment, Atherton Martin, resigned following the decision by Dominica's delegate to the International Whaling Commission to vote against the establishment of a South Pacific whaling sanctuary, in defiance of a cabinet decision to abstain.

1 October 2000: Douglas died of a heart attack, and the then Minister of Communications and Works, Pierre Charles, replaced him as Prime Minister.

December 2000: Legislation was passed making the crime of 'laundering' money punishable by imprisonment and a substantial fine.

6 March 2001: The DLP elected Pierre Charles as party leader.

July 2001: The Governments of Dominica and Japan agreed to promote bilateral co-operation in a broad range of areas.

The Dominican Republic

AD 650: The island which came to be known as Hispaniola (or Quisqueya, the Amerindian name), and which was later divided into the Dominican Republic and Haiti, was inhabited by the Taino.

1492: The Italian navigator in the service of the Spanish Crown, Cristoforo Colombo (Christopher Columbus), landed on the island and named it Hispaniola ('Little Spain'). Columbus founded a small settlement, Navidad, on Hispaniola.

1493: The first permanent European settlement, named Isabella, was founded on the island.

1496: The Spanish established the city of Santo Domingo.

1520s: The Spanish imported African slaves to replace the Amerindian labourers that had disappeared as a result of exploitation and illness.

1697: The western part of the island was ceded to France as Saint-Domingue (present day Haiti) by the Treaty of Ryswick. The remaining section, controlled by the Spanish, was called Santo Domingo.

1795: France gained nominal control of the whole island through the Treaty of Basle.

1808: The eastern side of Hispaniola was returned to Spanish rule.

1821: The Spanish administration became increasingly tyrannical, resulting in an uprising by the island's inhabitants, who proclaimed their independence.

1822: The President of Haiti, Jean-Pierre Boyer annexed Santo Domingo, and abolished slavery.

27 February 1844: Haitian rule was overthrown by a resistance group, La Trinitaria, under the leadership of Juan Pablo Duarte, and Santo Domingo declared its independence, forming the Dominican Republic.

1861: Owing to the threat of invasion by Haiti, President Pedro Santana requested the re-establishment of Spanish dominion.

1863–64: The Spanish Government withdrew its forces, following a popular revolt and intervention by the USA.

February 1865: Independence was regained following a 'War of Restoration'.

1882–99: The country came under the rule of Ulises Heureaux.

1905: The Republic's inability to pay its foreign creditors led to the establishment, by the USA, of a 50-year customs receivership.

29 November 1916–29 March 1924: The Dominican Republic was occupied by US military forces.

16 August 1930: The commander of the Dominican Army, Gen. Rafael Leónidas Trujillo Molina, seized power in a military *coup d'état*, overthrowing the elected President, Horacio Vásquez.

1937: Trujillo ordered the massacre of thousands of Haitian immigrants working on the country's sugar plantations.

16 May 1947: Trujillo was re-elected to the presidency with 92.9% of the votes cast, but ruled indirectly through his brother, Gen. Héctor Trujillo.

19 March 1950: An investigative committee appointed by the Council of the Organization of American States (OAS) found that Cuba and Guatemala had backed a 'Caribbean Legion', that aimed to invade the country and overthrow Trujillo. The OAS also censured the Dominican Republic for aiding Haitian exiles planning to overthrow the Haitian Government.

16 May 1950: Gen. Héctor Trujillo was re-elected to the presidency unopposed.

3 August 1960: President Trujillo resigned on grounds of ill-health and was replaced by his Vice-President, Dr Joaquín Balaguer Ricardo. Gen. Rafael Trujillo remained the absolute leader of the country. Following his appointment, Balaguer proposed that all political detainees and political exiles be granted a general amnesty.

20 August 1960: A meeting of OAS Foreign Ministers concluded that the Dominican Republic had supported a plot to assassinate the Venezuelan President. Almost every American nation subsequently severed diplomatic relations with the Dominican Republic.

30 May 1961: Gen. Rafael Trujillo was assassinated.

2 June 1961: Maj.-Gen. Rafael Trujillo, Jr, the eldest son of Gen. Rafael Trujillo, was appointed Chairman of the Joint Chiefs of Staff, thereby effectively taking control of the country. Balaguer remained as President.

23 October 1961: Following several days of severe rioting, Balaguer announced that Gen. Héctor Trujillo and Gen. José Trujillo, Gen. Rafael Trujillo's brothers, were to go into voluntary exile.

18 November 1961: In response to the decision by Héctor and José Trujillo to return from Bermuda to the Dominican Republic the USA warned that it would intervene to prevent any member of the Trujillo family from re-establishing control over the country.

19 November 1961: Maj.-Gen. Rafael Trujillo, Jr, resigned his position and announced that he was to leave the country. His two uncles and 29 other family members also went into exile in the USA.

27 November 1961: Opposition members demanded Balaguer's resignation and his replacement by a Council of State. However, Balaguer refused to stand down, leading to a general strike.

17 December 1961: Balaguer formed a Council of State, presided over by himself; the Council was to have full executive and legislative powers, and replaced the dissolved National Assembly.

4 January 1962: The OAS voted to lift economic and political sanctions on the Dominican Republic. Most member states subsequently restored diplomatic relations.

16 January 1962: A military *coup d'état*, led by Gen. Rodriguez Etchevarria, failed to overthrow the Council of State. Two days later Balaguer resigned as President and was replaced by Dr Rafael F. Bonnelly, one of the leaders of the National Civic Union. The Council of State was replaced by the Provisional Government.

20 December 1962: The leader of the Partido Revolucionario Dominicano (PRD), Juan Bosch Gaviño, was elected President at the country's first democratic elections for 38 years. The PRD also won majorities in the Chamber of Deputies and the Senate.

25 September 1963: The Government was deposed by a military *coup d'état*, led by Col (later Gen.) Elías Wessin y Wessin. Power was transferred to a three-man civilian junta. The USA and almost all other OAS members refused to recognize the new administration.

25–26 April 1965: The civilian junta was overthrown by military officers loyal to former President Bosch, led by Col Francisco Caamaño. A civil war broke out between pro-Bosch supporters and military units led by Wessin y Wessin. However, US troops were subsequently sent to the Dominican Republic and the violence was suppressed.

4 May 1965: Caamaño was sworn in as constitutional President.

8 May 1965: Caamaño's opponents established a three-man junta, the Government of National Reconstruction, under the leadership of Gen. Antonio Imbert.

24 May 1965: US troops came under OAS control and an OAS peace-keeping force was despatched to the Dominican Republic.

31 August 1965: The four-month civil war ended following the signature by the two sides of the Act of Dominican Reconciliation, which included a general amnesty for all those involved.

3 September 1965: The former Foreign Minister, Dr Héctor García-Godoy, was sworn in as President of the Provisional Government, under the terms of the Act of Dominican Reconciliation.

1 July 1966: Balaguer, the candidate for the Partido Reformista Social Cristiano (PRSC), was successful at presidential elections and was sworn in as President. The PRSC also secured a majority of seats in both houses of the new National Congress.

16 May 1970: Balaguer was re-elected as President. The PRD did not contest the elections, in protest at Balaguer's decision to seek a further term in office.

5 February 1973: A state of emergency was declared when a guerrilla force, led by Caamaño, landed in the Dominican Republic. It was defeated by the armed forces and Caamaño was killed in the fighting. The Government accused Bosch of planning the invasion, but he denied the charges and went into hiding.

16 May 1974: Balaguer was elected to his third consecutive term as President. The opposition did not contest the elections following a dispute over voting arrangements.

21 August 1974: The Senate approved a proposal whereby presidents were to be prevented from serving more than two consecutive terms in office.

8 June 1975: A group of Dominican exiles invaded the country with the aim of overthrowing Balaguer. The invasion was defeated and the group's leaders, Carlos Peña Jaquez and Claudio Caamaño (the nephew of Col Caamaño) subsequently received long prison sentences.

16 May 1978: The PRD candidate, Silvestre Antonio Guzmán Fernández, defeated former Presidents Balaguer and Bosch at presidential elections.

8 September 1978: Guzmán introduced an amnesty for political prisoners and exiles. The President also forced a large number of senior police and army officers to resign as part of his plan to depoliticize the security forces.

16 May 1982: The PRD candidate, Dr Salvador Jorge Blanco, defeated former Presidents Balaguer and Bosch at presidential elections.

3 July 1982: Guzmán committed suicide and was replaced as President on an interim basis by the then Vice President Jacobo Majluta, until Blanco took office as President, in August.

23–25 April 1984: Serious rioting took place at which some 50 people were killed, in protest at the Government's economic austerity programme.

16 May 1986: At presidential elections Balaguer of the PRSC narrowly defeated Majluta (PRD) and Bosch (representing the Partido de Liberación Dominicana—PLD). At simultaneous legislative elections, the PRSC won 21 of the 30 seats in the Senate and 56 of the 120 seats in the Chamber of Deputies.

January 1987: Former President Blanco was charged with corruption; he was later sentenced *in absentia* to 20 years' imprisonment.

February 1988: Six people were killed in riots that took place throughout the country in protest at the increases in the cost of living.

June 1989: A general strike took place, paralyzing the country for two days; furthermore, four people were killed and over 3,000 arrested at demonstrations held in protest at the deteriorating living conditions on the island. Protesters requested a significant increase in the minimum wage, a decline in prices and improved public utilities.

16 May 1990: Balaguer narrowly defeated Bosch at presidential elections, although he was not confirmed as President until two months following his election, following investigations into allegations of electoral fraud by the PRSC and the Junta Central Electoral (JCE). The PRSC also retained control of the Senate, securing 16 of the 30 seats, although it lost its majority in the Chamber of Deputies, obtaining 42 of the 120 seats.

16 May 1994: Balaguer was re-elected President for a two-year term after narrowly defeating José Francisco Peña Gómez, the leader of the PRD. Peña alleged serious irregularities in the conduct of the elections and the PRD boycotted Balaguer's inauguration. At legislative elections, held concurrently, the alliance of the PRD and Unidad Democrática (UD) secured 15 seats, while the PRSC won 14 and the PLD one seat. In the Chamber of Deputies the PRD/UD obtained 57 seats, while the PRSC gained 50 and the PLD 13 seats.

August 1994: Following discussions aimed at resolving the political crisis that ensued from the elections in May, a Pact for Democracy was signed. Under the terms of the agreement new elections were to be held and a series of constitutional reforms were to be enacted (including prohibiting the re-election of the President to a consecutive term). However, later that month the PRD withdrew from the legislature in protest at the decision by PRSC deputies to extend Balaguer's mandate. Following the approval by Congress of the terms of the Pact, Balaguer was confirmed as President.

30 June 1996: Following an inconclusive first round of presidential elections in May, a second round was organized, at which the PLD candidate, Leonel Fernández Reyna, was declared the winner, securing 51.3% of the votes, and defeating Peña Gómez who suggested that the authorities had conspired against him.

1997: Violent demonstrations and strikes took place throughout the year in protest at the deteriorating public services and high cost of living.

16 May 1998: At a general election the PRD won a majority in both legislative houses, securing 83 of a possible 149 seats in the Chamber of Deputies, and 24 seats in the Senate. The PLD obtained 49 and four seats, and the PRSC 17 and two seats.

16 May 2000: The PRD candidate, Rafael Hipólito Mejía Domínguez, defeated the PLD candidate, Daniel Medina Sánchez, and former President Balaguer at presidential elections. Mejía secured 49.9% of the vote, compared with 24.9% for the PRD candidate and 24.6% for Balaguer. Mejía was declared President, despite his failure to secure the constitutionally required 50% of the vote.

February 2001: A strike was called by public health workers, in protest at welfare reforms proposed by the Government.

1 November 2001: Former President Bosch died.

Ecuador

c. AD **1480:** The Inca civilization dominated various indigenous ethnic groups in the region.

1487: Quito was annexed by the Incas. The city was developed into a major administrative and military centre by the Inca Emperor, Huayna Capac.

1526: The Spanish explorer, Bartolomé Ruiz, landed on the coast.

1532: Francisco Pizarro, the leader of the Spanish force that conquered the Inca Empire, invaded the country.

1534: The city of Quito was captured by Sebastián de Belalcázar.

1537: Guayaquil was founded by Francisco de Orellana.

December 1540: Francisco Pizarro appointed his brother, Gonzalo Pizarro, Governor of Quito, as the colony was named.

1541–9 April 1548: Gonzalo Pizarro led a rebellion against Spanish rule of the colony. Spanish dominance was restored in 1548 when Pizarro was defeated and later executed by the Spanish army at Jaquijaguana. Quito was henceforth administered by the Viceroyalty of Peru.

1563: The Audiencia de Quito was established; the body advised the Viceroy and also functioned as a court of law.

1717–23: The Audiencia de Quito was administered by the Viceroyalty of New Granada in Bogotá (now Santafé de Bogotá, Colombia).

1724–38: The Audiencia de Quito became subject to the authority of the Viceroyalty of Peru.

1739: The Audiencia was returned to the authority of the Viceroyalty of New Granada.

1809: Revolts against Spanish rule were suppressed.

1822: Gen. José Antonio de Sucre, a Venezuelan officer of the forces of Simón Bolívar, defeated the Spanish army at the Battle of Pichincha, thus securing the territory's independence from Spanish rule. Ecuador subsequently became part of Bolívar's Federation of Gran Colombia, losing territory to Peru and Brazil.

May 1830: Ecuador seceded from Gran Colombia and attained independence as the Republic of Ecuador; Gen. Juan José Flores became the country's first President. Also at this time, the Galápagos Islands were annexed and a series of border wars with Peru began.

1833: Civil war arose between the Partido Conservador of Quito (PC, Conservatives), and the Partido Liberal, later Partido Liberal Radical of Guayaquil (PL, Liberals). This was the first in a long series of violent confrontations that dominated Ecuador's political climate in the 19th and 20th centuries, with the result of great political instability—by 1980 the country had had 19 constitutions and 88 heads of state.

1861–65: The Conservative Gabriel García Moreno took office as President. He supported the interests of the Roman Catholic Church, and concentrated on improving the country's infrastructure through a campaign of public works; a railway link between Quito and Guayaquil was built.

1869: García Moreno was re-elected to the presidency.

June 1895: Liberal forces revolted against the Government and seized power, led by Gen. Eloy Alfaro. During his term of office, which ended in 1901, Ecuador entered a period of modernization; railway lines were constructed and foreign investment encouraged.

1904: A border war against Peru took place.

1906: Gen. Alfaro was re-elected to the presidency. A new Constitution was introduced.

1916: Conflict regarding disputed territory in the border area resumed.

1924: The spread of disease affecting the country's major export commodity, cocoa, severely damaged the country's foreign-earning capacity; a recession ensued (exacerbated by the global economic depression of the 1930s) and the market for Ecuadorean cocoa never recovered.

1931–48: In a period of particular political instability, Ecuador had 21 governments, none of which served its full term of office.

1934: Dr. José María Velasco Ibarra was elected President. During the course of the mid-20th century, Velasco was elected President on five occasions and was overthrown four times.

1941: As a result of defeat in a further border war against Peru, Ecuador was forced to cede one province on the Pacific coast and substantial areas of land in the east.

1942: The Government was obliged by the USA to sign the Protocol of Rio de Janeiro, which recognised Peru's gains; however, Ecuador continued to claim sovereignty over the lost territories.

1944: Velasco was re-elected to the presidency. He was overthrown in 1947.

1948: Dr. Galo Plaza Lasso was elected President; during a period of relative political and economic stability he was able to complete his term of office (the first President to do so since 1924). The cultivation of the banana was developed as a new export crop.

1952: Velasco was elected President for the third time. in the same year, the waters around the Galápagos Islands were declared part of the exclusive maritime zone claimed by Ecuador, Chile and Peru.

1956: Camilo Ponce Enríquez defeated Velasco in the presidential election, becoming the first Conservative to be elected to the presidency since the revolution of 1895.

1960: Velasco returned to the presidency as a non-party Liberal.

1961: Velasco was deposed by a *coup d'état*. Dr. Carlos Julio Arosemena Monroy, hitherto the Vice-President, succeeded him.

July 1963: Monroy was deposed by a military junta led by Capt. (later Rear-Adm.) Ramón Castro Jijón. Despite its authoritarian nature, the regime gained an amount of initial popularity by enacting a programme of agrarian reform, the need for which had been a long-standing cause of tension in the country.

1966: After dismissing the Junta, The High Command of the Armed Forces appointed the former Minister of Economics, Clemente Yerovi Indaburu, as President.

October 1966: A Constituent Assembly was established.

November 1966: A new Constitution was enacted by the Constituent Assembly forbidding the intervention of the armed forces in elective politics. In consequence, Yerovi (who retained significant military support) was forced to resign. He was replaced by Dr Otto Arosemena Gómez.

1968: Velasco returned from exile to win the presidency for a fifth time.

1970: Amid a financial crisis caused by a decline in prices for the country's export commodities, President Velasco, supported by the military, suspended the Constitution and began rule by decree.

1971: Two US fishing boats were apprehended and fined after they were accused by the authorities of sailing unlawfully in Ecuadorean territorial waters; the incident marked the beginning of a so-called 'tuna war' between the two countries.

February 1972: Velasco was deposed by a military *coup d'état* led by Gen. Guillermo Rodríguez Lara, who became head of state. The new 'Revolutionary Nationalist Government,' which favoured state intervention in the economy, introduced a new agrarian reform by promoting the establishment of agricultural co-operatives. Following the discovery of substantial petroleum reserves in the Oriente region, the country entered a period of unprecedented prosperity.

1973: The Galápagos archipelago became a province, with its capital at Puerto Baquerizo Moreno.

October 1973: The country joined the Organization of the Petroleum Exporting Countries (OPEC).

1976: Rodríguez Lara transferred power to a military junta under the leadership of Vice-Adm. Alfredo Poveda Burbano. The new junta announced its intention to restore democracy.

January 1978: A new Constitution was approved in a national referendum.

July 1978: The first round of voting in presidential elections failed to produce an outright winner; a second round was subsequently scheduled.

April 1979: Jaime Roldós Aguilera, from the Concentración de Fuerzas Populares (CFP), was elected President, with over 68% of the second-round votes cast.

August 1979: Roldós took office and the new Constitution entered into force. The new President promised social justice and economic development, and guaranteed freedom for the press; however, his efforts at introducing reforms were obstructed by Congress (in which opposition groups held a majority) and the trade unions.

January 1981: A dispute with Peru over the sovereignty of the Cordillera del Cóndor escalated into a further border conflict. A cease-fire was declared some days later under the auspices of the guarantors of the Rio Protocol of 1942 (Argentina, Brazil, Chile and the USA). Ecuador had refused to recognize the Protocol as it awarded the area, which affords access to the Amazon River system, to Peru.

May 1981: Roldós was killed in an air crash and was succeeded as President by Dr Osvaldo Hurtado Larrea. Amid a deteriorating economic situation, Hurtado's efforts to reduce government spending were opposed by both left- and right-wing politicians and by trade-union leaders.

January 1982: President Hurtado's attempts to reach an amicable solution with Peru regarding the countries' border dispute proved unpopular within the armed forces: the chiefs of each of the services resigned and the Minister of Defence was dismissed.

December 1982–January 1983: Further conflict with Peru occurred.

March 1983: The introduction of a series of austerity measures encountered direct opposition from the trade unions and private-sector employees.

January 1984: In a presidential election, no candidate won an overall majority and a second round was scheduled. In concurrent congressional elections, the ruling Democracia Popular—Unión Demócrata Cristiana (DP—UDC) suffered substantial losses.

May 1984: In the second round of voting in the presidential elections. León Febres Cordero, the leader of the Partido Social Cristiano (PSC) and candidate of the conservative Frente de Reconstrucción Nacional, obtained 52.2% of the votes cast, narrowly defeating Dr Rodrigo Borja Cevallos, of the left-wing Izquierda Democrática (ID). Febres introduced a number of populist reforms and began a programme of economic liberalization, including the reduction if state intervention in the economy and the promotion of export trade.

1986: Lt-Gen. Frank Vargas Pazzos, the Chief of Staff of the Armed Forces, was dismissed and detained, after he made allegations of corruption against the Minister of Defence and the Commander of the Army (both of whom subsequently resigned). Vargas, who had initially barricaded himself and his supporters in his base, then organized a second rebellion at the military base where he had been detained, before being captured by government forces.

June 1986: In congressional elections, President Febres Cordero's coalition lost its majority; a concurrent referendum on constitutional amendments proposed by the President also resulted in defeat. The decline in petroleum prices had necessitated further measures of economic austerity, which were proving highly unpopular.

January 1987: President Febres Cordero was abducted by rebel troops demanding the release of Vargas. Vargas was granted an amnesty in exchange for the President's release.

March 1987: An earthquake at Napo killed several hundred people and destroyed some Trans-Amazonian petroleum pipelines.

July 1987: A total of 58 members of the Air Force were sentenced to up to 16 years' imprisonment for their involvement in the abduction of the President.

May 1988: Borja won the presidential election, defeating Abdalá Bucaram Ortiz of the Partido Roldosista Ecuatoriano (PRE) in the second round of voting with 46% of the ballot, compared with Bucaram's 41%.

August 1988: Borja took office and introduced a programme of moderate social reforms known as *gradualismo*. However, a concurrent programme of economic austerity provoked large public demonstrations.

October 1988: The guerrilla organization, Montoneros Patria Libre (MPL), proposed the establishment of dialogue between the Government and the numerous armed rebel organizations operating in the country.

October 1988: The President of the Supreme Court of Justice was assassinated in Quito.

February 1989: The Government began a campaign to confiscate weapons belonging to paramilitary organizations.

March 1989: The guerrilla organization Alfaro Vive ¡Carajo! (AVC) urged paramilitary groups to surrender their weapons. In return, the civil rights of AVC members were guaranteed and the Government promised to initiate a dialogue in return for the group's demobilization. The MPL continued its violent opposition.

late September 1989: Ecuador signed a preliminary agreement with the International Monetary Fund (IMF), which was expected to provide funding to the value of US $1,250m.

October 1989: A plot to overthrow the President and replace him with the Vice-President was foiled. Radical right-wing groups, led by Febres Cordero, were subsequently identified as the conspirators.

January 1990: Febres Cordero was arrested and charged with the embezzlement of public funds during his tenure as President.

May 1990: An indigenous-rights group marched into Quito demanding official recognition of land rights for the indigenous population.

June 1990: In mid-term legislative elections, the ID and its allies lost control of Congress to an alliance of the PSC and the PRE.

June 1990: The Confederación Nacional de Indígenas del Ecuador (CONAIE) began an uprising covering seven Andean provinces. They demanded the return of traditional community-held lands, the recognition of Quechua as an official language and compensation by petroleum companies for environmental damage.

August 1990: Discussions between Government and CONAIE collapsed.

October 1990: Following the Government's loss of its congressional majority, confrontational politics caused a legislative *impasse*. Dr Averroes Bucaram, the President of Congress, impeached several ministers, who were subsequently dismissed by Congress. Congress then dismissed the Supreme Court justices and other high-ranking members of the judiciary, and appointed new courts. Both the Government and the judiciary refused to recognize these actions, on the grounds that Congress had exceeded its constitutional powers. Bucaram then announced that Congress would initiate impeachment proceedings against Borja himself. The move was averted when three opposition deputies transferred their allegiance, restoring Borja's congressional majority; Bucaram was dismissed as President of Congress.

January 1991: The Frente Unitario de Trabajadores (FUT) and CONAIE organized a joint anti-Government protest against the Government's decision on the level of increase of the minimum monthly wage.

February 1991: Dialogue between Government and CONAIE resumed after the seizure by indigenous groups of eight petroleum wells. The AVC concluded the demobilization process, which had begun in February 1989.

August 1991: Further border clashes occurred between Ecuador and Peru.

September 1991: The British Embassy in Quito was occupied by eight members of a dissident faction of the AVC. The dissidents demanded the liberation of their imprisoned leader.

October 1991: The AVC was absorbed by the ID.

January 1992: Discussions with Peru regarding the border dispute continued.

May 1992: At legislative elections, the PSC gained the greatest number of seats in the National Congress (21 of 77); the PRE secured only 13. The recently formed Partido Unidad Republicano (PUR) led by former PSC presidential candidate, Sexto Durán Ballén, won 12 seats, the ID seven seats and the PC six seats. The PUR and PC governed in alliance.

May 1992: Indigenous groups were granted legal title to more than 1m. ha of land in the province of Pastaza.

5 July 1992: Durán defeated Jaime Nebot Saadi of the PSC in the second round of the presidential election, gaining 58% of the ballot.

September 1992: The Government announced austerity measures to strengthen the economy. The restructuring of the public sector prompted violent demonstrations and several bomb attacks in Quito and Guayaquil.

late November 1992: Ecuador withdrew from OPEC, claiming that the costs of affiliation were excessive and the refusal to increase Ecuador's production quota was detrimental to the country's economic prospects.

May 1993: A general strike took place.

August 1993: A 'Modernization Law', which allowed the privatization of state-owned companies, was approved by Congress.

November 1993: The Consejo Nacional de Modernización (CONAM) was formed and the country's entry into the Andean free-trade area was implemented. Teachers began a strike demanding wage increases and the reform of the education system.

January 1994: The Government's decision to increase the price of fuel by more than 70% provoked violent demonstrations and a general strike. Indigenous groups also intensified their protests.

February 1994: The increase in the price of fuel was declared unconstitutional by the Tribunal of Constitutional Guarantees. President Durán refused to recognize the ruling.

May 1994: Mid-term congressional elections took place: the PSC was victorious; the PUR-PC governing alliance won only nine of the 77 seats.

June 1994: The indigenous movement organised large-scale demonstrations and a general strike in protest at the Land Development Law, which allowed for the commercialization of Amerindian lands for farming and extraction. The military was mobilized and a state of emergency declared. Seven protesters were killed. The law was later declared unconstitutional by the Tribunal of Constitutional Guarantees, although President Durán refused to accept the ruling.

July 1994: The Land Development Law was modified to extend the rights of landowners and those employed to work on the land.

28 August 1994: Following low levels of voter participation in a national referendum on constitutional reform, only one of the eight proposed reforms was approved.

January 1995: Protests against the Government's economic programme of austerity measures and privatizations resulted in the deaths of two students in Quito.

January 1995: Reports of Peruvian incursions into Ecuadorian territory resulted in a serious confrontation between the two countries. Following offers from the Organization of American States (OAS) and the four guarantor nations of the Rio de Janeiro Protocol, delegations from the two countries met for negotiations in Rio de Janeiro.

February 1995: Ecuador and Peru agreed to a cease-fire.

March 1995: The Minister of Finance resigned.

May 1995: The FUT began a national strike against the Government's introduction of 'corrective' economic measures introduced to attenuate the impact of the financial crisis caused by war against Peru. Ecuador recognised the validity of the Rio de Janeiro Protocol, and thus Peru's claim to the disputed territory, for the first time.

July 1995: Ecuador and Peru agreed on the delimitation of the demilitarized zone in the disputed area.

July 1995: A serious political crisis emerged when Vice-President Alberto Dahik admitted giving funds to opposition deputies in return for their support for the Government's economic programme. Following his refusal to resign, impeachment procedures were initiated against Dahik. The President of the Supreme Court and two other justices were dismissed. In addition, petroleum workers initiated a strike in protest at the impact of government policy on the petroleum industry.

September 1995: Impeachment proceedings were begun against the new Minister of Finance, Mauricio Pinto. The Superintendent of Banks resigned his post, following accusations that he had attempted to hinder the case against Dahik in the Supreme Court. In addition, troops were sent to the Galápagos Islands to suppress disturbances among the islanders, who were demanding increased political and financial autonomy.

October 1995: Petroleum workers resumed strike action and occupied the seat of the state petroleum company, taking two cabinet ministers as hostages. The affair ended with the resignation of the Minister of Energy and Mines at the end of the month.

2 October 1995: Impeachment proceedings against Dahik began.

11 October 1995: Following an appeal by President Durán, Dahik resigned. The former Minister of Education, Dr Eduardo Peña Triviño, was subsequently elected as Vice-President by Congress. A further two cabinet ministers resigned in support of Dahik.

13 October 1995: The entire Cabinet resigned in order that a reorganization of portfolios could take place.

November 1995: The electorate rejected all government proposals for constitutional change in a referendum. The death of a student in a strike organized by teachers and students resulted in the resignation of the Minister of Education.

January 1996: Army units were deployed at prominent sites throughout the country to prevent further unrest. The Government began formal negotiations aimed at agreeing a final solution to the border problems with Peru.

March 1996: An industrial dispute by transport workers resulted in serious disruption in Quito and provoked a series of strikes in other sectors.

May 1996: Presidential elections failed to produce an overall majority for any candidate. Freddy Ehlers, the candidate of the newly formed Movimiento Nuevo País-Pachakutik, which represented indigenous and environmental groups, secured 21% of the votes cast in the presidential election and was the third most popular candidate becoming an important force in Congress.

July 1996: Abdalá Bucaram Ortiz of the PRE unexpectedly won the second round of voting in the presidential election, receiving 54.5% of the ballot compared with the 45.5% obtained by Nebot Saadi of the PSC.

August 1996: Bucaram took office and appointed a team of businessmen to advise him on economic policy.

September 1996: In a scandal at the Ministry of Finance, seven senior officials were arrested on charges of embezzlement (estimated to involve more than US $300m). Bucaram appeared vindicated in his anti-corruption policies.

October 1996: The Government and Peru signed the Santiago Agreement, which was to provide a framework for a definitive solution on the border issue.

January 1997: Increases of up to 600% in the price of certain commodities as well as growing concerns at Bucaram's unconventional leadership resulted in a two-day general strike.

February 1997: An estimated 2m. demonstrators marched through the streets of Quito demanding President Bucaram's resignation. The President responded by declaring a national holiday and a one-week closure of schools across the country, and stated his support for the strike. Troops were deployed throughout the capital as violent clashes erupted between protesters and security personnel. President Bucaram was barricaded inside the presidential palace.

6 February 1997: At an emergency session, Congress voted to dismiss the President on the grounds of mental incapacity and a state of emergency was declared. Fabián Alarcón Rivera, who had been appointed acting President, urged demonstrators to storm the presidential palace—despite this, Bucaram refused to leave office. Rosalía Arteaga, the Vice-President, claimed to be the legitimate constitutional successor to Bucaram, leading to fears of military intervention to resolve the situation.

9 February 1997: Bucaram fled from the presidential palace to Panama, where he successfully obtained political asylum.

10 February 1997: Arteaga was declared President after narrowly winning a congressional vote.

11 February 1997: Arteaga resigned and Alarcón was reinstated as President. He announced a reorganization of cabinet portfolios (in which no members of the two largest parties in Congress, the PSC and the PRE, were included) and the creation of a commission to investigate allegations of corruption against Bucaram and his administration.

March 1997: The Government requested Bucaram's extradition from Panama on charges of misappropriating some US $90m of government funds. Ecuador produced an official complaint against the Peruvian army's use of landmines along the border.

April 1997: The President of the Supreme Court declared that Bucaram's extradition would only be possible once a prison sentence had been issued.

May 1997: Bucaram announced his intention to seek election to the presidency. Congress imposed an indefinite ban on Bucaram's candidacy in any future presidential election.

25 May 1997: A national referendum on electoral reform, the modernization of the judiciary and the authenticity of Alarcón's position was held. The vote revealed support for both Bucaram's removal from office and Alarcón's presidency.

June 1997: Support for Alarcón was jeopardized by the launch of an official inquiry into allegations that leading drugs-traffickers in the country had contributed to political party funds, and particularly to Alarcón's Frente Radical Alfarista.

July 1997: In response to the recent referendum's approval of the depoliticization of the judiciary, Congress dismissed all 31 judges of the Supreme Court. The President of the Supreme Court condemned the action as unconstitutional.

August 1997: The announcement that a National Assembly to review the Constitution would not be convened until August 1998 resulted in a two-day strike, which was supported mostly by the indigenous population. A former FRA colleague filed charges of embezzlement against President Alarcón in the Supreme Court.

September 1997: The Government announced that elections for the 70 representatives to the National Assembly would take place in late November. A total of 11 indigenous organizations convened a mass rally in Quito to establish guidelines for their own constitution.

January 1998: The Supreme Court issued a prison sentence against Bucaram, after he was convicted *in absentia* of slandering two political rivals.

January 1998: Congress approved a law aiming to preserve the Galápagos Islands' unique environment. A section of the law, which provided for an extension of the marine reserve around the island from 15 to 40 nautical miles, attracted criticism from powerful fishing interests in the country and was consequently vetoed by President Alarcón. He was subsequently criticized for having placed commercial pressures at a higher priority than the regional environment.

February 1998: The National Constituent Assembly announced a series of institutional reforms: Congress was to be enlarged from 82 to 121 seats and the presidential term extended from four years to four years, five months and five days. Mid-term elections were abolished and it was established that the Vice-President would complete a presidential term in the case of the indefinite absence of the President.

31 May 1998: In the first round of voting in presidential elections, Jamil Mahuad Witt, the DP candidate and Mayor of Quito, emerged as the strongest contestant. Alvaro Novoa Pontón of the PRE came second, followed by ex-President Borja and Freddy Ehlers. In concurrent elections to the enlarged Congress, the DP won 32 seats, the PSC 27 seats, the PRE 24 and the ID 18.

12 July 1998: A second round of voting took place; Mahuad narrowly defeated Novoa, winning 51.2% of the votes cast.

10 August 1998: Mahuad took office; his new Government contained mainly independent members. Upon his inauguration, the new Constitution came into force.

September 1998: Significant increases in the cost of domestic gas and electricity, public-transport fares and fuel prices, occurred.

October 1998: A general strike was organized by FUT and CONAIE to protest against the price increases. Violent confrontations resulted in the deaths of four people.

26 October 1998: Ecuador and Peru signed an accord, which confirmed Peru's claim regarding the delineation of the border. Ecuador was accorded navigation and trading rights on the Amazon and the opportunity to establish two trading centres in Peru. Moreover, Ecuador was granted 1 sq km of territory, as private property, at Tiwintza in Peru, where many Ecuadorian soldiers killed during the conflict of 1995 were buried.

February 1999: The Minister of Finance, Fidel Jaramillo Buendia, resigned over disagreement concerning economic policy. Fuel shortages prompted the resignation of the Minister of Energy and Mines, Patricio Rivadeneira García. The leader of the left-wing Movimiento Popular Democrático and member of Congress, Jaime Hurtado González, was assassinated in Quito; the killing was linked to a right-wing Colombian paramilitary group. In the same month, the former Interim President, Alarcón, was arrested on charges of illegally hiring personnel during his term in office as Speaker of Congress.

March 1999: Ecuador's military presence was strengthened at its border with Colombia after reports of Colombian paramilitary troops in the country. A substantial decrease in the value of the sucre led President Mahuad to declare a week's national holiday to reduce the pressure on the currency. A 60-day state of emergency was subsequently declared by the Government in response to a two-day strike, organized by trade unions and indigenous groups. The economic crisis deepened with the collapse of the Banco del Progreso.

April 1999: An emergency plan entitled 'Ecuador 2000', which included a number of measures required to gain the support of the IMF, was revealed by the Government.

late April 1999: Ecuador, four other Latin American countries and the USA submitted a complaint to the World Trade Organization (WTO) that the European Union (EU) unfairly favoured Caribbean banana producers (chiefly in the former European colonies there). The WTO upheld the complaint and the USA imposed trade sanctions against certain EU goods.

May 1999: Tiwintza was officially transferred from Peru to Ecuador.

June 1999: Four new members of the board of the central bank were appointed in a plan approved by the IMF. The Government hoped that the IMF would provide an estimated US $400m. in initial funding.

5 July 1999: Public-transport workers began a strike in protest at increases in fuel prices and the new economic plan proposed by the IMF. Other trade-unions and indigenous groups joined the strike. The Government declared a state of emergency. An armed confrontation between protestors and soldiers in the central Andes occurred when an estimated 2000 indigenous-rights demonstrators took possession of the communications antennae in Ambato.

14 July 1999: Congress repealed the state of emergency. President Mahuad deemed that the country was not sufficiently stable to warrant such action, however, and reimposed it a few hours later.

16 July 1999: Prresident Mahuad agreed to abandon any further increase in fuel prizes and to allow co-operatives and transport companies to access 'frozen' bank accounts. The state of emergency was lifted.

August 1999: The country experienced further financial instability. Unable to meet its foreign-debt repayments, the Government cancelled its payments for a period of 30 days, in an attempt to gain an agreement on restructuring (Ecuador's external debt totalled more than US $13,000m).

September 1999: The Minister of Finance, Alicia Lucia Armijos, resigned.

September 1999: Colombian guerrillas were suspected of abducting 12 foreign nationals, several of whom were employees of a petroleum company.

November 1999: EU restrictions on Latin American banana exporters were eased.

January 2000: A state of emergency was declared as a result of increasing unrest amid indications that the economic situation was deteriorating.

9 January 2000: Despite previous objections, President Mahuad announced a decision to adopt the US dollar as the currency of Ecuador. The President of the Central bank resigned, claiming that the proposed conversion rate of the sucre was unfeasible.

21 January 2000: Mahuad was forced to flee from the presidential palace, following large-scale protests by indigenous-rights demonstrators in Quito, supported by sections of the armed forces; some of the protesters occupied Congress. Antonio Vargas, the President of CONAIE, Gen. Carlos Mendoza, the Chief of Staff of the Armed Forces, and Carlos Solorzano, the former President of the Supreme Court, established a new 'parliament of salvation'. The next day, Gen. Mendoza dissolved the triumvirate and appointed a former Vice-President, Gustavo Noboa Bejerano, as President. Indigenous-rights groups continued to demonstrate against the appointment of Noboa and the dismissal of the triumvirate.

27 January 2000: President Noboa appointed Pedro Pinto Rubianes as the new Vice-President.

February 2000: Noboa formed a Cabinet, comprising members of the DP, the PRE and the PSC. Heinz Moeller Freile became Minister of Foreign Affairs, and Francisco Huerta Montalvo became the new Minister of the Interior. The Ley de Transformación Económica, which included 'dollarization', fiscal reform and an expanded policy of privatization, was approved by Congress. CONAIE organised demonstrations against the adoption of the dollar and called for a referendum to dissolve Congress and the Supreme Court. It also demanded amnesty for troops involved in the *coup d'état* of January.

April 2000: The IMF confirmed a one-year stand-by agreement, conditionally approving US $2,000m in multilateral aid over the next two years. The Government formally began the process of replacing the sucre with the dollar. Dollarization resulted in a stabilization of the exchange rate and a restoration of popular confidence in the economy. Concessions to opposition groups successfully diminished popular discontent, although CONAIE continued to call for mass action against privatizations and dollarization.

May 2000: The entire military high command was replaced and an amnesty was declared for those arrested in connection with the *coup d'état* which deposed Mahuad. The Government's announcement of an IMF-approved economic adjustment package was opposed by trade unions and indigenous groups and resulted in the resignation of the Minister of Finance and Public Credit, Jorge Guzmán Ortega. Both the DP and PSC performed well in the municipal and provincial elections, suggesting significant popular acceptance of structural reforms.

August 2000: Relations between PD and PSC came under increasing strain when the independent Susana González was controversially elected legislative President for the remainder of the session. The Government offered to resign in support of the President after the governing coalition lost its majority in Congress.

September 2000: CONAIE renewed negotiations with the Government.

September 2000: Ecuador began to receive Colombian refugees as violence escalated in that country.

December 2000: Two attacks on the trans-Ecuadorian petroleum pipeline resulted in at least 16 fatalities and 41 injuries.

December 2000: The Minister of Finance and Public Credit, Luis Yturralde, who was opposed to the economic measures favoured by the IMF, resigned.

January 2001: Increases in fuel prices and transport costs were met with popular resentment. Indigenous-rights protesters occupied Quito and roadblocks were erected across the Andean highlands and the Amazon lowlands.

late January 2001: It was feared that an environmental catastrophe had occurred after a petroleum tanker ran aground less than one kilometre from the Galapagos Islands. Some 190,000 gallons of petroleum leaked into the sea, although it was thought that favourable weather conditions had alleviated the environmental damage.

February 2001: Following the intensification of protests, the Government imposed a national state of emergency, which was lifted only after it acceded to the demands for a reduction in fuel prices. Moeller announced his intention to ask the US Government for financial aid to support the creation of an economic and military 'buffer' zone between Ecuador and Colombia.

April 2001: Continued congressional opposition to certain of the conditions imposed by the IMF for the release of its credits, most notably the increase in Value Added Tax (VAT).

4 May 2001: Congress approved an increase in VAT; the IMF subsequently agreed to release funds.

El Salvador

***c.* 3000 BC:** The nomadic Nahua people, subsequently incorporating various sub-groups, including the Pipil, the Pocomames and the Lencas, migrated into Central America.

AD 31 May 1522: The Spanish Adm. Andrés Niño led an expedition to Central America from the colony of New Spain (Mexico) and disembarked on the Isla Meanguera, in the Gulf of Fonseca.

June 1524: The Spanish Lt Pedro de Alvarado entered Cuscatlán (now El Salvador), but was forced to retreat by the Pipil.

1525–28: Alvarado brought Cuscatlán under Spanish rule, within the Captaincy-General of Guatemala.

5 November 1811: An indigenous fight for independence, organized by Fr José Matías Delgado and centred in San Salvador, began.

15 September 1821: The Central American provinces, including El Salvador, declared their independence from Spain.

June 1822: El Salvador resisted a Central American union with Mexico.

1823: The United Provinces of Central America, comprising Costa Rica, El Salvador, Guatemala, Honduras and Nicaragua, was formed, under the rule of Gen. Manuel José Arce.

12 June 1824: El Salvador proclaimed the first Constitution of Central America.

1833: An Indian rebellion, led by Anastasio Aquino, failed to alter the system of land ownership in El Salvador.

1838: Following the collapse of the United Provinces of Central America, El Salvador achieved full political independence.

1 February 1841: The first Constitution of El Salvador came into effect.

mid-19th century: A small group of local landowners and merchants transformed the country into a specialized producer of agricultural exports.

1860s: Following the collapse of the indigo market, the production of coffee replaced that of more traditional crops.

1882: Common land was abolished and three-quarters of all land passed into the private ownership of an élite of landowners and merchants (known as the 'Fourteen Families') comprising only about 2% of El Salvador's inhabitants.

1907: Costa Rica, El Salvador, Guatemala, Honduras and Nicaragua signed a Treaty of Peace and Friendship.

1931: Following a *coup d'état*, Gen. Maximiliano Hernández Martínez became President, replacing Arturo Araújo, who had been elected earlier in the year.

1932: President Hernández suppressed a peasant uprising, led by the Partido Comunista Salvadoreño (PCS) under the leadership of Agustín Farabundo Martí; it was alleged that 30000 people were killed (including Farabundo Martí).

1944: President Hernández was deposed, and was replaced by Gen. Salvador Castañeda Castro.

14 December 1948: President Castañeda Castro was overthrown in a military *coup d'état*, to be replaced by a Provisional Council of Government, headed by Lt-Col Oscar Osorio.

26–28 March 1950: Lt-Col Osorio was elected President.

1951: The Organization of Central American States was founded by Costa Rica, El Salvador, Guatemala, Honduras, Nicaragua and Panama.

4 March 1956: Lt-Col José María Lemus was elected President.

8 January 1959: The Central American Common Market, comprising El Salvador, Guatemala, Honduras, Nicaragua, and later Costa Rica, came into force.

26 October 1960: Following a bloodless *coup d'état*, President Lemus was replaced by a military junta.

25 January 1961: A *coup d'état*, headed by Col Aníbal Portillo, replaced the junta with a new Military-Civilian directorate.

December 1961: In legislative elections the recently established conservative Partido de Conciliación Nacional (PCN) won all 54 seats in the Constituent Assembly.

16 January 1962: The Assembly, renamed the Asamblea Nacional (National Assembly), assumed power over the directorate. A new Constitution came into effect, providing for a republican, democratic and representative form of government, composed of three Powers—Legislative, Executive and Judicial—which would operate independently.

29 April 1962: Col Julio Adalberto Rivera, of the PCN, was elected unopposed to the presidency.

5 March 1967: The PCN candidate, Col Fidel Sánchez Hernández, was elected to the presidency.

1969: ORDEN, the first of a series of extreme right-wing terrorist groups, or 'death squads', was established; these groups engaged in intimidation and murder of those who supported political reform.

June 1969: Ill-feeling concerning the treatment of football supporters during three World Cup qualifying round matches between the El Salvador and Honduras national teams resulted in the countries terminating diplomatic relations. About 12,000 Salvadorean migrants resident in Honduras were expelled from the latter country, and a state of emergency was declared in El Salvador.

3–18 July 1969: Military skirmishes between Salvadorean and Honduran armed forces culminated, on 14 July, in Salvadorean troops advancing some distance into Honduras. After the intervention of the Organization of American States (OAS), a cease-fire was agreed on 18 July. A final peace treaty was not signed until 1980, and the underlying territorial dispute, which was referred to the International Court of Justice (ICJ) in 1986, was not resolved until 1992. About 2000 people were killed in the conflict, which is sometimes known as the 'Football War'.

1970: Cayetano Carpio, the Secretary-General of the PCS, broke from the party in order to pursue a campaign of armed insurrection.

8 March 1970: The PCN won 38 National Assembly seats in a general election, the remaining 16 seats being won by the Partido Demócrata Cristiana (PDC).

20 February 1972: Col Arturo Armando Molina Barraza was elected President. In legislative elections held in the following month, the PCN once again won a majority, with 38 seats; the Unión Nacional de Oposición (in which the PDC was the major coalition partner) won seven seats. Violent protests and an unsuccessful *coup d'état* followed allegations of massive electoral fraud, resulting in some 100 deaths; the leader of the PDC, José Napoleón Duarte, was exiled.

20 February 1977: The PCN candidate, Gen. Carlos Humberto Romero Mena, was elected President.

15 October 1979: A military *coup d'état*, headed by Col Jaime Abdul Gutiérrez and Col Adolfo Arnoldo Majano, forced President Romero into exile.

January 1980: An unstable coalition Government of military officers and members of the PDC was formed.

24 March 1980: The Roman Catholic Archbishop of San Salvador, Oscar Arnulfo Romero y Galdames, an outspoken supporter of human rights, was assassinated.

April 1980: The Frente Farabundo Martí para la Liberación Nacional (FMLN) guerrilla front was formed from various extreme left-wing groups, marking the commencement of a more intensive civil war. The FMLN also had a political wing, the Frente Democrático Revolucionario (FDR). The Salvadorean Government refused to recognize the FDR while it was linked with the guerrillas.

30 October 1980: The Governments of El Salvador and Honduras signed a peace treaty.

13 December 1980: José Napoleón Duarte was sworn in as the first civilian President since 1931.

30 September 1981: A new right-wing political party, Alianza Republicana Nacionalista (ARENA), was formed. It was to be alleged that various ARENA politicians had links with the right-wing 'death squads'.

28 March 1982: In elections to the new National Constituent Assembly no party obtained an absolute majority. The five right-wing parties together obtained 60% of the total votes, and so formed a Government of National Unity. Maj. Roberto D'Aubuisson Arrieta, leader of ARENA, became President of the Assembly.

29 April 1982: A politically independent banker, Dr Alvaro Magaña Borja, was accepted by all parties as interim President.

20 December 1983: The new Constitution entered into effect, providing for a republican, democratic and representative form of government, composed of the three independent Powers. Parts of the 1962 Constitution had been in abeyance since the *coup d'état* of 1979.

1 July 1984: In the second round of voting in presidential elections, Duarte was re-elected President, defeating D'Aubuisson.

31 March 1985: At legislative and municipal elections the PDC secured a clear majority in the National Assembly.

1 October 1986: The conflicting territorial claims of Honduras and El Salvador over three islands in the Gulf of Fonseca and a small area of land on the common border were submitted to the International Court of Justice (ICJ) for arbitration.

October 1987: In discussions between the Government and the FDR-FMLN, an agreement was reached on the formation of two committees to study the possibility of a cease-fire and an amnesty.

November 1987: The Partido Social Demócrata (PSD), the Movimiento Nacional Revolucionario (MNR) and the Movimiento Popular Social Cristiano (MPSC) united to form a left-wing alliance, the Convergencia Democrática (CD). The MNR and MPSC were, however, to remain as members of the FDR-FMLN.

20 March 1988: At legislative and municipal elections ARENA secured control of more than 200 municipalities, including San Salvador. The party secured 31 seats in the National Assembly, compared with the PDC's 23 seats.

19 March 1989: Alfredo Félix Cristiani Burkard of ARENA was elected President of El Salvador. The FMLN had advocated a boycott of the election and intensified its campaign of violence, resulting in the deaths of more than 40 people on election day alone. The level of abstention was estimated at almost 50%.

7 November 1989: The UN Security Council authorized the creation of the UN Observer Group for Central America (ONUCA), a multinational military force, to monitor developments in the region.

11 November 1989: The FMLN launched a military offensive.

16 November 1989: The head of a San Salvador Jesuit university, and five other Jesuit priests, were killed by gunmen.

12 January 1990: The FMLN announced that it would accept an offer made by the Salvadorean Government whereby the UN Secretary-General, Javier Pérez de Cuéllar, was to arrange the reopening of peace talks.

November 1990: A renewed FMLN offensive was undertaken by the newly-proclaimed National Army for Democracy (the establishment of which marked the reorganization of the FMLN's previous divisions into a more conventional army structure).

10 March 1991: In elections to the National Assembly (enlarged from 60 to 84 seats) ARENA lost its majority, but continued to command considerable support, with 44.3% of the votes and 39 seats. The party also retained significant support in the local elections, with victories in 175 of the 262 municipalities.

1 July 1991: The UN Security Council created an observer mission to El Salvador (ONUSAL), charged with the verification of accords reached between the Government and the FMLN.

December 1991: The heads of state of Costa Rica, El Salvador, Guatemala, Honduras, Nicaragua and Panama signed the Protocol of Tegucigalpa to the agreement establishing the Organization of Central American States, thus founding the Central American Integration System.

December 1991: A new peace initiative was announced, incorporating a formal cease-fire on 1 February 1992, under the supervision of some 1,000 UN personnel.

16 January 1992: The formal peace accord was ratified at Chapultepec Castle in Mexico City.

September 1992: El Salvador and Honduras accepted the ruling of the ICJ regarding disputed territory; Honduras was awarded two-thirds of the disputed mainland territory, but El Salvador received two of the three contested islands in the Gulf of Fonseca. A resolution of the issue, which had led to numerous disputes between the two countries, was finally achieved in January 1998, when the parties signed a convention specifying the rights and obligations of those affected, including the right to choose between Honduran and Salvadorean citizenship.

November 1992: In accordance with the terms of the December 1991 peace accord, the Comisión de la Verdad (Truth Commission) announced the names of more than 200 military personnel alleged to have participated in abuses of human rights during the civil war. Various legal proceedings followed.

15 December 1992: The conflict was formally concluded and the FMLN was officially registered and recognized as a legitimate political party.

20 March 1994: Presidential, legislative and municipal elections took place. ARENA's presidential candidate, Armando Calderón Sol, won in the second round of voting, while ARENA candidates also achieved considerable success in the legislative elections (retaining 39 seats in the Assembly) and in the municipal poll (securing an estimated 200 municipalities). FMLN candidates won 21 seats in the Assembly and a number of rural municipalities.

6–12 December 1994: The Resistencia Nacional (RN) and the Expresión Renovadora del Pueblo (ERP—formerly the Fuerzas Armadas de la Resistencia Nacional and the Ejército Revolucionario Popular guerrilla groups, respectively) announced their withdrawal from the FMLN, owing to a divergence of political interests.

28 March 1995: The formation of a new centre-left political force, the Partido Demócrata (PD), comprising the ERP, the RN, the MNR and a dissident faction of the PDC, was announced.

April 1995: The ONUSAL mandate was extended until the end of April 1995, and was then replaced by a small contingent of UN observers, MINUSAL, which remained in El Salvador until the formal termination of the mission on 31 December 1996.

10 October 1996: The National Assembly approved a constitutional amendment whereby the death penalty for civilians (abolished in 1971) was to be reinstated for a number of offences including rape, abduction and aggravated murder.

16 March 1997: In municipal and legislative elections the FMLN gained a significant increase in support, while support for ARENA declined. ARENA won 28 National Assembly Seats, and the FMLN 27; however, divisions within the latter prevented them from becoming a coherent alternative to ARENA nationally. ARENA won control in 161 municipalities, and the FMLN in 48.

19 May 1997: A new Ministry of the Environment and Natural Resources was created to address the country's serious environmental problems.

7 March 1999: The presidential election was won by Francisco Flores Pérez of ARENA, who obtained 52.0% of the votes, thereby obviating the need for a second round of voting. Voter participation was the lowest recorded in the country's history, with more than 60% of the 3.2m. registered voters choosing to abstain.

12 March 2000: In legislative elections the FMLN became the largest single party in the National Assembly, winning 31 of the 84 seats. ARENA secured 29 seats, followed by the PCN (14 seats), the PDC (five seats), the Centro Democrático Unido (CDU—an electoral alliance comprising the CD and the PSD, three seats) and the Partido Acción Nacional (PAN, two seats). In concurrently-held municipal elections the FMLN gained control of 78 of the 262 municipalities, while ARENA won 127.

21 March 2000: The FMLN proposed a governability pact with ARENA. However, relations between the two parties immediately became strained.

1 May 2000: ARENA formed an alliance with the PCN and the PDC in order to prevent the FMLN claiming the presidency of the Assembly. Although the party gaining the most seats was, traditionally, given the presidency, in order to prevent a political *impasse* it was decided that the post was to be rotated annually among the three largest parties, the PCN, ARENA and the FMLN.

1 January 2001: In an attempt to stabilize the economy, lower interest rates, and stimulate domestic and foreign investment, the US dollar was introduced as an official currency alongside the colón, at a fixed rate of exchange.

13 January 2001: An earthquake, measuring 7.6 on the Richter scale, hit El Salvador, causing landslides in many areas. More than 700 people were killed, with many more unaccounted for, an estimated 4,000 people were injured, and a further 50000 people were left homeless. Total damages and losses were estimated at US $1,600m., representing around 10% of gross domestic product in 2000.

May 2001: The PCN supported the FMLN in granting US $1,000 each to 37,708 former village guards who had been recruited by the army during the civil war. However, this compensation was not covered in the peace accords and President Flores vetoed the bill, claiming that the country could not afford to pay compensation.

Grenada

c. **1000 BC:** Several South American tribes, including the Calvignoid, the Galibi, the Suazoids and the Caribs, settled in the area.

AD 15 August 1498: The Italian navigator in the service of the Spanish Crown, Cristoforo Colombo (Christopher Columbus) named the island Concepción, in honour of the Virgin Mary. The Spanish later renamed the island Granada, after the Spanish city.

1609: The British settled in Granada, establishing Megrin Town, but were soon forced to leave by the Caribs.

1639: The French colonized the island, which they renamed La Grenade, and successfully established a settlement.

1763: The island was ceded to the British under the Treaty of Paris, and renamed Grenada.

1779: The French re-captured the island, following discord between the British rulers and the French settlers.

1783: Grenada was returned to the British under the Treaty of Versailles.

1795: The French leader of a rebellion, Julien Fedon, temporarily gained control over the island. British sovereignty was swiftly restored.

1877: Grenada became a British Crown Colony.

1885: Grenada was administered as part of the Windward Islands. Each island was granted a Legislative Council.

2 August 1956: The British Parliament adopted a bill establishing the British Caribbean Federation, which was to have a joint Government, Legislature and Supreme Court.

1958: Grenada joined the West Indies Federation.

6 October 1959: The British Colonial Office announced the abolition of the post of Governor of the Windward Islands as part of the reform of the constitutions of the Leeward Islands. The reforms were designed to reduce the administrative differences between the members of the West Indies Federation.

27 March 1961: The Grenada United Labour Party (GULP) won eight of the 10 available seats in the Legislative Council and the GULP leader, George Clyne, was appointed Prime Minister.

27 February–5 March 1962: Representatives of Barbados and the seven Leeward and Windward Island colonies met in Barbados. They requested that the British Government approve a new federation, known as the Little Eight, in the event of the collapse of the West Indies Federation.

31 May 1962: The British Government formally dissolved the moribund West Indies Federation.

19 June 1962: The British authorities suspended Grenada's Constitution and dismissed the Chief Minister, Eric Gairy (later Sir Eric Gairy), and his Government after an inquiry revealed serious financial irregularities in the island's public finances.

September 1962: The Grenada National Party (GNP) won a majority at legislative elections, and the party's leader, Herbert Blaize, became Chief Minister.

1 November 1964: At a further meeting of the Little Eight, it was decided that the proposed federation would not include Grenada.

1 November 1966: Antigua, Barbados, Dominica, Grenada, Montserrat, St Christopher (St Kitts) and Nevis, St Lucia and St Vincent founded the West Indies (Associated States) Council of Ministers, which discussed proposals for a 'small island federation'. These talks, however, collapsed and the islands, with the exception of Montserrat, subsequently achieved full independence separately.

3 March 1967: Grenada was granted the status of an Associated State, and became self-governing, although the United Kingdom retained responsibility for defence and foreign relations.

27 February 1972: The GULP won 13 of the 15 seats available in legislative elections and Gairy was reappointed Premier. The party had campaigned on a platform of full independence from the United Kingdom.

25 July 1972: The Governments of Dominica, Grenada, Guyana, St Christopher and Nevis, St Lucia and St Vincent signed the Declaration of Grenada, which created the framework for an Eastern Caribbean political union.

7 February 1973: Grenada became the first of the Associated States to achieve full independence from the United Kingdom. The celebrations took place amid serious political violence between Government supporters and opponents.

7 December 1976: The GULP was returned to power with a reduced majority at general elections. The leftist New Jewel Movement (NJM) claimed to have evidence of serious electoral malpractice by Government agents.

4 January 1978: The acting Minister of Agriculture, Innocent Belmar, was assassinated. Belmar had had serious disagreements with Gairy, who was widely considered to have been responsible for the murder.

1 February 1978: The British Government announced that it was reconsidering its provision of aid to Grenada, after it was revealed that Gairy had expanded the armed forces and purchased significant amounts of military hardware from the military regime in Chile.

13 March 1979: The Government was overthrown in a *coup d'état* planned by the NJM. The NJM leader, Maurice Bishop, replaced Gairy as Prime Minister. A People's Revolutionary Government (PRG) was installed and the Constitution suspended. The PRG was subsequently recognized by the United Kingdom, Canada and the USA.

15 October 1979: Some 20 people were arrested and charged with planning to overthrow the PRG. The following month a second alleged *coup d'état* against the PRG was discovered; the plotters had allegedly received US assistance.

19 June 1980: Bishop and other leading PRG members narrowly escaped assassination when a bomb planted by a disaffected former army officer exploded at a public function.

11 July 1981: Four people were arrested by the police following an attempt to overthrow the PRG, allegedly with the support of the US Central Intelligence Agency (CIA).

27 July 1982: Bishop visited the USSR and signed a number of important co-operation agreements.

14 October 1984: Bishop dismissed the Deputy Prime Minister, Bernard Coard, following accusations that he had been preparing a *coup d'état* against the PRG. Coard reportedly enjoyed the support of the armed forces and of one faction of the PRG.

17 October 1984: The Commander of the Armed Forces, Gen. Hudson Austin, announced that Bishop had been arrested and forced to resign as Prime Minister. Several other ministers were also detained and foreign journalists were expelled from the island.

19 October–20 October 1984: Bishop and 10 associates were executed at an army base, and a number of pro-Bishop demonstrators were killed by the security forces. The following day Austin instituted a Revolutionary Military Council (RMC) to replace the PRG. The coup received widespread international condemnation and the majority of Grenada's regional neighbours severed diplomatic ties with the country.

25 October 1984: US forces launched an invasion, referred to as Operation Urgent Fury, to overthrow the RMC. They encountered serious resistance from local troops and from armed Cuban construction workers, and some 42 US soldiers and 100 Grenadians and Cubans were killed. The US invasion was strongly condemned internationally as unjustified. The RMC forces were defeated, while Coard and Austin were detained.

15 November 1984: An eight-member Advisory Council was sworn in by the Governor-General and the Constitution was restored. The Council was intended to act as an interim non-political Government.

22 November 1984: US troops relinquished control of the island to a peace-keeping force drawn from Caribbean states.

3 December 1984: A general election to the restored House of Representatives was won by the New National Party (NNP), a conservative grouping led by Herbert Blaize. The opposition dismissed the elections as a fraud perpetrated by foreign interests.

4 December 1986: Coard and 13 other associates were sentenced to death for their part in the execution of Bishop and 10 of his supporters.

17 February 1987: Blaize announced a major cabinet reshuffle amid rumours that several ministers were planning to remove him from office.

19 December 1989: Blaize died. The Deputy Prime Minister, Ben Jones, was appointed acting Prime Minister.

13 March 1990: At a general election, the centrist National Democratic Congress (NDC) won seven of the 15 seats in the House of Representatives, and the NDC leader, Nicholas Braithwaite, was appointed Prime Minister.

17 August 1994: Braithwaite, whose Government had been criticized for its reductions in public spending, resigned as leader of the NDC. He was subsequently replaced by the then Minister of Agriculture, George Brizan.

20 June 1995: The NNP defeated the NDC at legislative elections, winning eight of the 15 seats in the House of Representatives. The NNP leader, Keith Mitchell, became Prime Minister.

16 July 1997: The Minister of Health, Housing and the Environment, Grace Duncan, was dismissed after allegedly making insulting remarks about the Prime Minister.

29 November 1998: The Minister of Foreign Affairs, Raphael Fletcher, resigned his position and defected to the opposition GULP, leaving Mitchell with a minority in the House of Representatives.

18 January 1999: At anticipated general elections the NNP secured a comprehensive victory, receiving 62.2% of votes cast and obtaining all 15 seats in the House of Representatives. The NDC obtained 24.9% of votes, while the United Labour alliance received only 11.6%. Mitchell was reappointed Prime Minister.

26–27 June 2000: The Organisation for Economic Co-operation and Development (OECD) and the Financial Action Task Force (FATF, part of the OECD) criticized Grenada's financial system, naming it as a tax haven. The OECD requested that the country reform its rules by 31 December 2005 or face sanctions.

January 2001: The Minister of Foreign Affairs, Mark Isaac, was dismissed, reportedly following his decision to re-establish diplomatic relations with Libya without gaining prior approval from the Cabinet. He was succeeded by Elvin Nimrod.

13 March 2001: The Government announced the closure of 17 offshore banks following pressure from the OECD.

25 June 2001: The Government cancelled the licences of a further six offshore banks.

Guatemala

***c.* 2000 BC:** Mayan fishing and farming villages emerged along Guatemala's Pacific coast.

300–800: Mayan civilization thrived in the highlands of what is now Guatemala.

***c.* 400:** The Mayans began to build large temples; a number of kingdoms emerged.

AD 1523: Pedro de Alvarado conquered Guatemala in the name of the Spanish Crown.

1524: The first capital, Santiago de los Caballeros, was founded at Iximché.

1776: The capital was moved to present day location of Guatemala City.

1821: Guatemala achieved independence from Spain, but the following year was incorporated into Mexico.

1823: Following a revolution in Mexico, Guatemala became a founding member of the United Provinces of Central America.

1825: Manuel José Arce became the first President of the federation of the United Provinces, based in Guatemala City. However, other Central American states revolted against his policies, which included the abolition of slavery.

1829: The Honduran Gen. Francisco Morazán defeated the federal forces and captured Guatemala City. He ruled by decree and encouraged the expansion of trade and industry to encourage settlement.

1830: Morazán declared himself President of the federation.

1838: Following several uprisings and many years of war and unrest, the federation was dissolved.

3 December 1839: Guatemala became an independent republic, with Mariano Rivera Paz as its President.

14 May 1844: José Rafael Carrera Turcios was elected President and ruled until 1865, except for a short period in 1848–49. The conservative Carrera upheld the power of the ruling classes and restored the dominant position of the Roman Catholic Church.

21 October 1854: Carrera Turcios was appointed President-for-life.

24 May 1865: Following the death of Carrera Turcios in April, Vicente Cerna Sandoval was elected President.

4 June 1873: Justo Rufino Barrios Auyón was elected President and instigated substantial liberal reforms including the disestablishment of the Roman Catholic Church and the modernization of the economy.

2 April 1885: Barrios Auyón was killed in battle and the following year Barillas Bercián was elected President.

8 February 1898: José María Reina Barrios, who had been appointed President in 1892, was assassinated, and was succeeded by Manuel José Estrada Cabrera.

15 April 1920: President Estrada Cabrera was declared unfit to rule by Congress and driven into exile after being ousted by a *coup d'état*. Carlos Herrera y Luna replaced him as President.

14 February 1931: Gen. Jorge Ubico siezed power.

4 July 1944: Following strikes and a period of civil unrest, Ubico was overthrown in a *coup d'état* and Juan Frederico Ponce Vaides was appointed acting President.

15 March 1945: Guatemala's first democratic elections were held, at which Juan José Arévalo Bermejo was elected President.

1948: Guatemala became a member of the UN.

15 March 1951: Arbenz Guzmán was elected President.

27 June 1954: Arbenz Guzmán was overthrown, with the support of the USA, by 'liberation' forces made up of Guatemalan exiles and led by Col Carlos Castillo Armas. Castillo Armas assumed power in September.

1957: Castillo Armas was assassinated by one of his guards.

2 March 1958: Gen. Miguel Ydigoras Fuentes was elected President.

31 March 1963: Ydigoras Fuentes was overthrown in a *coup d'état* led by Col Enrique Peralta Azurdia. Peralta Azurdia assumed full powers as Chief of Government, suspended the Constitution and dissolved the legislature.

1964: A Constituent Assembly was elected.

1965: The Constituent Assembly introduced a new Constitution.

1 July 1966: Dr Julio César Méndez Montenegro was elected President.

1 July 1970: The candidate for the Movimiento de Liberación Nacional (MLN), Col (later Gen.) Carlos Manuel Araña Osorio was elected President.

2 July 1974: Gen. Kjell Laugerud García of the MLN was elected President, amid allegations of electoral fraud.

1977: The USA suspended sales of arms and the provision of economic and military aid to Guatemala.

1 July 1978: Gen. Fernando Romeo Lucas García was elected President, amid allegations of electoral fraud.

1980: Guerrilla activity intensified and, following an incident at the Spanish embassy which led to the deaths of some 36 people, Spain severed diplomatic relations with Guatemala.

1982: The three leading guerrilla organizations united to form the Unidad Revolucionaria Nacional Guatemalteca (URNG) in the belief that they could seize power. The counter-insurgency campaign escalated with the establishment of community-based paramilitary groups called Patrullas de Autodefensa Civil (PAC), which assisted the army in killing indigenous communities thought to be supporters of the guerilla movement.

7 March 1982: Gen. Angel Aníbal Guevara was elected President, amid allegations of electoral fraud. He was prevented from taking office by a *coup d'état* held later that month. The army installed a three-man military junta with Gen. José Efraín Ríos Montt as leader. The Congreso Nacional was closed, and the Constitution and political parties suspended.

June 1982: Gen. Ríos Montt dissolved the Junta and assumed the presidency.

July 1982: After unsuccessful attempts to negotiate with the guerilla forces, Gen. Ríos Montt declared a state of siege and imposed censorship of the press. Gen. Ríos Montt's power was threatened throughout 1982 by several attempted *coups d'état*, which he managed to forestall.

March 1983: The army was implicated in the massacre of 300 indigenous agricultural workers. Gen. Ríos Montt declared a 30-day amnesty for guerrillas and political exiles and lifted the state of siege. He also announced the formation of an electoral tribunal to oversee the transfer of power from military rule to civilian government.

8 August 1983: Gen. Oscar Humberto Mejía Victores seized power in a *coup d'état*, pledging a return to democracy. He abolished the secret tribunals and ended press censorship.

November 1983: In response to alleged human-rights violations by the Government, and following the murder of six US aid-workers in northern Guatemala, the US Congress banned all aid to Guatemala.

July 1984: Elections were held to a Constituent Assembly, which later drafted a new Constitution.

May 1985: The new Constitution was promulgated.

November 1985: Legislative and presidential elections were held concurrently. Mario Vinicio Cerezo Arévalo, of the Partido de la Democracía Cristiana de Guatemala (PDCG) received 38.7% of the votes cast. However, as he did not obtain the requisite majority, a second round of voting was held in December, at which Cerezo Arévalo was elected President, receiving 68.4% of the valid votes cast. The PDCG formed the majority party in the new Congreso Nacional.

October 1987: Government and URNG representatives met in Spain to discuss peace proposals. The negotiations, however, ended without agreement.

May 1988: President Cerezo Arévalo survived an unsuccessful *coup d'état* and postponed further negotiations with the URNG.

July 1988: President Cerezo Arévalo rejected the URNG's proposal for a truce after a plot to carry out a *coup d'état* was uncovered.

January 1989: A new leftist group emerged: the Comando Urbano Revolucionario joined the URNG guerrillas. President Cerezo Arévalo refused to negotiate with the URNG for as long as its members remained armed.

May 1989: Another attempted *coup d'état* was suppressed.

August 1989: A secret right-wing military organization perpetrated a series of terrorist attacks in an attempt to destabilize the Government.

December 1989: A major counter-insurgency operation was launched by the Government. President Cerezo Arévalo accused the ruling party in El Salvador of supplying weapons to the right-wing 'death squads' of Guatemala.

March 1990: Preliminary negotiations between the URNG and the Commission of National Reconciliation began in Oslo, Norway.

14 January 1991: Jorge Serrano Elías, of the Movimiento de Acción y Solidaridad (MAS), was elected President, receiving 68.1% of the valid votes cast in the second ballot. Serrano Elías formed a minority Government with the support, initially, of the Partido de Avanzada Nacional (PAN).

April 1991: A fresh round of negotiations between the URNG and the Government was begun in Mexico City. The meeting resulted in an agreement on negotiating procedures and an agenda for further negotiations. However, in an attempt to destabilize the Government's efforts at national reconciliation, members of the state security forces, believed to be acting independently of their superiors, launched a campaign of violence.

May 1992: Fears of a possible military *coup d'état* were raised by a series of bombings and bomb threats in the capital. The campaign was widely recognized as the activity of right-wing elements of the military, which aimed to force the Government to abandon its plans to implement tax reforms and to cease negotiations with the URNG.

July 1992: The Minister of the Interior, Fernando Hurtado Prem, was forced to resign amid allegations of police brutality.

25 May 1993: With the support of the military, Serrano Elías partially suspended the Constitution and dissolved the Congreso Nacional announcing his intention to rule by decree. The USA, the European Union (EU) and several other countries introduced economic sanctions.

1 June 1993: International pressure and overwhelming domestic opposition led the military to effect a return to constitutional rule, forcing the resignation of Serrano Elías. The Minister of National Defence, Gen. José Domingo García Samayoa, assumed control of the country, pending the election of a new president. An attempt by Vice-President Gustavo Adolfo Espina Salguero to assume the presidency was prevented by a legislative boycott of his ratification. He was later ruled ineligible for the post by the Constitutional Court.

3 June 1993: The entire Cabinet, excluding García Samayoa and the Minister of the Interior, Francisco Perdomo Sandoval, resigned. On the same day, charges, including violation of the Constitution, abuse of authority and embezzlement, were presented against Serrano, Espina and Perdomo.

5 June 1993: Following an order by the Constitutional Court, Congress reconvened and elected Ramiro de León Carpio to complete what remained of Serrano's term. The URNG announced a unilateral cease-fire as a gesture of goodwill to the incoming President. The USA subsequently restored its aid programme to Guatemala.

August 1993: President de León Carpio requested the voluntary resignation of the legislature and the Supreme Court as an initial measure in a campaign to eradicate corruption from state institutions and restore public confidence in his Government.

September 1993: The army announced that it would be resuming military operations against the rebels, which had been suspended in June.

October 1993: The Government presented a revised peace plan to the UN, providing for the creation of a Permanent Forum for Peace and renewed cease-fire negotiations; the plan was, however, rejected by the rebels.

January 1994: Negotiations between the Government and the URNG were resumed in Mexico, and resulted in an agreement on the resumption of formal peace negotiations.

March 1994: President de León Carpio oversaw the signing of a human-rights accord with the URNG, which included a mandate for a UN mission, the Human Rights Verification Mission in Guatemala (MINUGUA), which was formally established in September to supervise the implementation of the accord.

August 1994: Legislative elections were held at which Ríos Montt's Frente Republicano Guatemalteco (FRG) won the greatest number of seats (32 of a total of 80), followed by PAN and the PDCG. Also in that month the URNG withdrew from the peace negotiations and accused the Government of failing to observe the human rights provisions agreed in March.

November 1994: Peace negotiations between the Government and the URNG resumed, but failed to make any progress.

February 1995: A new timetable for negotiations between the Government and the URNG was announced. The new agenda provided for a cease-fire agreement by June and the signing of a definitive peace accord in August. However, the negotiations continued beyond the agreed deadline, and throughout 1995, without agreement on the substantive issues.

September 1995: In an attempt to lend impetus to the peace process, the Government announced the demobilization of the paramilitary force, the Comisionados Militares.

7 January 1996: Legislative and presidential elections were held and were won, respectively, by the PAN and its leader, Alvaro Enrique Arzú Irigoyen, who received 51.2% of the valid votes cast at the second ballot. His main opponent, the FRG's candidate Alfonso Portillo Cabrera, obtained 48.8% at the second ballot.

March 1996: A temporary, cease-fire was announced by the guerrillas.

19 September 1996: An agreement was reached by the Government and the URNG, providing for constitutional changes and legislation that would democratize the role of the military and intelligence services by making them subordinate to elected bodies.

29 December 1996: The Government and the URNG signed the Agreement for a Firm and Lasting Peace, which consisted of 11 peace accords.

March 1997: The URNG demobilized, under UN observation.

June 1997: The URNG registered as a political party in formation.

1998: The PAC demobilized, as did the Policia Ambulante Militar (PAM), and a new civilian police force was created.

28 April 1998: The auxiliary bishop of the metropolitan diocese of Guatemala City, Juan José Gerardi Conedera, was assassinated. Two days earlier Gerardi had presented a comprehensive report documenting atrocities committed during the civil war, of which army personnel were found to be responsible for some 80%.

October 1998: The Congreso Nacional approved a series of constitutional reforms advocated in the 1996 peace accords, concerning the rights of indigenous peoples, the role of the armed forces and the police and the strengthening of the courts. The changes required ratification by a referendum, which was scheduled to be held in 1999.

25 February 1999: The Comisión para el Esclarecimiento Historico, established to investigate the events of the civil war, published a report on the extent of the atrocities committed. It found that some 200,000 people had been killed or had 'disappeared' between 1962 and 1996, the vast majority of them Mayan Indians. It attributed responsibility for some 90% of these crimes to state military and paramilitary terrorism.

16 May 1999: A referendum was held to ratify the constitutional changes approved by the Congreso Nacional the previous year; the proposals were rejected.

7 November 1999: Legislative and presidential elections were held. The FRG won 63 of the 113 seats in the Congreso Nacional, while PAN secured only 37. The Alianza Nueva Nación (ANN), a left-wing grouping including the URNG, obtained nine congressional seats, while three minority parties secured the remaining four seats. At the first round of voting in the presidential election, Portillo Cabrera, the candidate for the FRG failed to obtain the required 50% of the votes cast, necessitating a second round in December, at which Portillo Cabrera won 68.3% of the valid votes cast, defeating Oscar Berger Perdomo of PAN, and was duly elected President.

2 December 1999: The Guatemalan peace activist Rigoberta Menchú filed a case before the Spanish courts accusing Ríos Montt and the former military rulers Gen. Romeo Lucas García and Gen. Oscar Humberto Mejía Victores, as well as other officers, of crimes of genocide, state terrorism and torture. The case was rejected by the courts the following year, on the grounds that there was no impediment to their trial within the Guatemalan judicial system.

14 January 2000: The new Government of President Portillo Cabrera took office, and was notable for its inclusion of former PAN, PDCG and URNG political leaders, as well as several prominent human-rights activists. Furthermore, civilians were appointed to key posts. The replacement of the EMP with a civilian security force also began, although it did not proceed smoothly.

April 2000: The Government admitted responsibility for human-rights violations at the annual meeting of the UN High Commission for Human Rights (UNHCHR).

June 2000: It was revealed that FRG deputies had illegally amended alcohol-tax legislation after it had been passed by the Congreso Nacional; legal proceedings began against Ríos Montt and 23 FRG deputies the following year, although they were later exonerated.

August 2000: The Government accepted responsibility for 44 of 150 cases of human-rights violations being studied by the Inter-American Commission on Human Rights (IACHR, part of the Organization of American States—OAS).

December 2000: The IACHR ruled that the military was responsible for the torture and murder of guerilla leader Efrain Bamaca Valásquez, following his capture in 1992. Also in December, the ODHA filed a case against the President of the legislature, Ríos Montt, for human-rights violations during his tenure as President.

February 2001: Rumours of a *coup d'état* were reported, prompting a visit by Secretary-General of the OAS, César Gavira Trujillo, to demonstrate support for President Portillo Cabrera.

6 June 2001: A number of indigenous communities filed genocide charges against Ríos Montt and other senior officers in his command, for his role in a series of massacres by the army during his presidency.

8 June 2001: The former intelligence chief, Col (retd) Disrael Lima Estrada, his son, Capt. Byron Lima Oliva, and a former member of the presidential guard, José Obdulio Villanueva, were convicted of the murder of Bishop Gerardi in 1998; a priest, Mario Orantes, was convicted of conspiring in his death.

Guyana

pre-AD 900: Amerindian groups, including the Warrau, Arawak and Carib, settled in the area of the Guianas (derived from an Amerindian word meaning 'land of many waters').

1499: Spanish explorers first charted the territory.

***c.* 1595:** The legend of the empire of El Dorado, which was believed to be in the area of the Guianas, began.

1598: The Dutch made their first voyage to the area.

1616: The Dutch founded a settlement at Kyk-Over-Al, an island in the Essequibo river. The Dutch Westindische Compagnie (West India Company—WIC, founded in 1621) later took control of the colony.

1627: A second colony was founded at the mouth of the Berbice river by the WIC.

1665: England declared war on the Netherlands and, in the following year, the Fort Kyk-Over-Al settlement was captured. The Dutch regained control of the fort in 1667.

***c.* 1680:** The WIC promulgated what is thought to have been the country's first Constitution, giving the Commander of the Essequibo colony complete administrative authority over the land, assisted by a Council consisting of the Sergeant of the Garrison and the captains of the ships that were then in port.

1732: The Berbice region established a Constitution, which instituted a Governor, a Council of Government and two Councils of Justice (criminal and civil).

1741: A third colony, at the mouth of the Demerara river, was founded. The three communities cultivated tobacco and sugar, and African slaves were imported to work on the plantations.

1763–64: The Berbice revolt began on 23 February 1763 when slaves in the colony began to protest at their inhumane treatment.

1781: The British took control of the three colonies, establishing Fort St George.

1782: The colonies fell into the hands of the French, who constructed a new town, Longchamps, on the site of Fort St George.

1784: The colonies were restored to the Dutch, who renamed the town of Longchamps as Stabroek. It later became the capital. Following the dissolution of the WIC, owing to financial difficulties, the colonies came under the control of the Dutch Government.

1792: The United Colony of Demerara and Essequibo was established.

1796: In renewed warfare, the colonies were taken by the British. They were returned to Dutch rule on 25 March 1802, under the Treaty of Amiens, but once again came under British control in the following year.

1812: The town of Stabroek was renamed George Town, in honour of the British Monarch, George IV.

1814: The territories were finally ceded to Britain by the Dutch, under the Treaty of Vienna.

18–20 August 1823: Some 200 slaves were killed in a revolt on the east coast of Demerara.

1831: Sir Benjamin D'Urban was appointed the first Governor of British Guiana.

1834: Slavery was abolished under the Emancipation Act. It was replaced by an apprenticeship system, for which the former slaves were paid wages. However, the system was abolished in 1838. Subsequently, both indentured and immigrant labourers were brought from China, Germany, India and Portugal.

1837: George Town had its first Mayor and town council.

1838: British Guiana was divided into three counties: Demerara, Essequibo and Berbice.

1842: George Town became the city of Georgetown.

1848: The first railway in South America, extending from Georgetown to the east coast of Demerara (and subsequently further extended), opened. It was built by the Demerara Railway Company. The Company was sold to the Government in 1922, and the railway finally closed following independence.

1886: Following the discovery of gold in 1857, the first mining regulations were issued. Diamond deposits were discovered in 1890 and manganese ore was found in 1903.

3 October 1899: The Tribunal of Arbitration, established to resolve a border dispute between British Guiana and Venezuela, rejected the latter's suggestion of a boundary on the Essequibo river. Venezuela had claimed 130,000 sq km (50,000 sq miles) of land west of the Essequibo river, based on a papal bull of 1493, referring to Spanish colonial possessions.

1928: A new Constitution entered into effect, with British Guiana becoming a Crown Colony under the administration of a Governor-General. A Legislative Council was established, comprising members appointed by the British Colonial Office. Although women were given the vote, suffrage remained limited by income.

1928–36: A series of protests and strikes occurred, as a result of economic and social conditions in British Guiana.

27 November 1947: In general elections the British Guiana Labour Party (BGLP) won six of the 14 seats in the Legislative Council.

24 April 1953: In the first general election under universal adult suffrage (introduced in 1952), the left-wing People's Progressive Party (PPP), led by Dr Cheddi Bharat Jagan, won 18 of the 24 seats on the Legislative Council.

9 October 1953: The British Government suspended the Constitution, claiming that the country was under threat of Communist subversion.

4 January 1954: An interim administration was appointed, comprising a Speaker, three *ex-officio* members and 24 nominated members.

1957: Following the split of the PPP in 1955, former members founded the People's National Congress (PNC), under the leadership of Forbes Burnham.

August 1957: In a general election the PPP won nine of the 14 seats in the Legislative Council, while the PNC won three seats. Dr Jagan became Chief Minister.

1958: The United Democratic Party (UDP, formerly the National Democratic Party, founded in 1953) merged with the PNC.

July 1961: A new Constitution, providing for internal self-government, was adopted.

August 1961: In a general election the PPP won 20 of the 35 seats in the Legislative Assembly, and Dr Jagan was again appointed premier.

1962: Venezuela renewed its claim to territory west of the Essequibo river. The Port of Spain Protocol of 1970 put the issue in abeyance until 1982.

18 April–6 July 1963: Widespread disturbances and a general strike followed an attempt by the Government to introduce new labour-relations legislation.

23 October 1963: In an attempt to resolve a political *impasse*, a proportional representation electoral system was implemented.

14 December 1964: In a general election the PPP won 24 of the 53 seats in the Legislative Assembly, while the PNC won 22 seats and the right-wing The United Front (TUF) obtained seven seats. A coalition Government was formed by the PNC and TUF, with Burnham appointed as premier.

December 1965: Antigua, Barbados and British Guiana signed an agreement to establish the Caribbean Free Trade Area (CARIFTA), which entered into effect (with the additional participation of Trinidad and Tobago) on 1 May 1968.

26 May 1966: The colony of British Guiana became an independent State, within the British Commonwealth, and was renamed Guyana. Burnham was appointed the first Prime Minister and the Governor-General was Sir Richard Luyt.

1968: Despite the resignation of TUF ministers within the PNC-TUF coalition before a general election, the PNC obtained the majority of votes and Burnham continued as Prime Minister.

23 February 1970: A new Constitution renamed the country the Co-operative Republic of Guyana. The office of Governor-General was abolished and, in March, Arthur Chung was elected non-executive President.

1973: CARIFTA became CARICOM (the Caribbean Community and Common Market), a grouping primarily of British Commonwealth Caribbean countries.

1973: The PNC was victorious in general elections, although the results of the poll were disputed by the opposition parties.

1974: In the Declaration of Sophia (named after a Georgetown suburb), the PNC committed itself to stronger socialist policies, and announced that government institutions were arms of the party.

1976: Arthur Chung was re-elected non-executive President.

July 1978: Following a referendum that gave the Assembly power to amend the Constitution, elections to the Assembly were postponed, and the legislature was to draft a new Constitution.

18 November 1978: About 900 members of a religious cult, the People's Temple, committed mass suicide, following the murder of US Congressman Leo Ryan and some members of his party, who were investigating allegations that members were being held against their will. The community, later known as Jonestown, had been established by a US citizen, James Warren Jones, in Guyana in 1974.

June 1980: Internal opposition to the Government increased after the assassination of Dr Walter Rodney, leader of the Working People's Alliance (WPA, formed in 1973). The Government was widely believed to have been involved in the incident; however, in 1988 an official inquest finally produced a verdict of death by misadventure.

6 October 1980: The new Constitution entered into force, under which the non-executive presidency was abolished and the head of government assumed the post of executive President. Guyana was also divided into ten regions, each with a regional democratic council.

December 1980: All opposition parties except the PPP and TUF boycotted the elections to the National Assembly. In the election the PNC won 41 of the 53 seats, although allegations of substantial electoral malpractice were made. Burnham was formally inaugurated as executive President in January 1981.

1983: Guyana and Venezuela referred the territorial dispute over 130,000 sq km of Guyanese territory to the west of the Essequibo river to the United Nations.

August 1985: Burnham died, and was succeeded as President by Desmond Hoyte, hitherto the First Vice-President and Prime Minister.

December 1985: In a general election the PNC won 78% of the votes cast and 42 of the elective seats in the National Assembly; Hoyte continued as executive President.

January 1986: Five opposition parties formed the Patriotic Coalition for Democracy (PCD).

January 1990: A movement for legal and constitutional change, Guyanese Action for Reform and Democracy (GUARD), initiated a series of mass protests, urging the Government to accelerate the process of democratic reform.

January 1991: The forthcoming general election was postponed in order to compile a new electoral register. The legislative term of the Assembly was further extended in March, May and July, owing to continuing problems relating to electoral reform. The publication of the revised electoral register in September revealed widespread inaccuracies, including the omission of an estimated 100,000 eligible voters. The election was re-scheduled for mid-December.

28 November 1991: Hoyte declared a state of emergency, legitimizing a further postponement of the election. In mid-December the state of emergency was extended until June 1992. A revised electoral register was finally approved by the Elections Commission in August.

5 October 1992: The general election resulted in a narrow victory for the PPP in alliance with the CIVIC movement (a social and political movement of businessmen and professionals), which secured 32 of the 65 elective seats in the National Assembly, while the PNC secured 31 seats. The end of the PNC's 28-year period in government provoked riots in Georgetown, in which two people were killed.

9 October 1992: Dr Jagan took office as President, appointing Samuel Hinds as Prime Minister.

August 1994: In the first municipal and local-government elections to be held since 1970, the PPP/CIVIC alliance secured control of 49 of the 71 localities concerned.

19 August 1995: A major environmental incident was caused by the spillage of some 3.5m. cu m of cyanide-tainted water from a breach in a tailings pond (a reservoir where residue from the gold extraction process is stored), into the Omai river, a tributary of the Essequibo river. Omai Gold Mine Ltd was closed for an indefinite period, pending an inquiry into the incident; however, it was permitted to resume operations in February 1996.

6 March 1997: Following the death of Dr Jagan, Hinds succeeded to the Presidency. He subsequently appointed Janet Jagan, the widow of the former President, to the post of Prime Minister.

15 December 1997: At the general election a series of delays in the verification of votes prompted accusations of electoral fraud. With some 90% of the votes counted, the Chairman of the Elections Commission declared Janet Jagan President. According to the final election results, released on 31 December, the PPP/CIVIC alliance secured 34 seats, the PNC won 26, and TUF, the Alliance for Guyana and the Guyana Democratic Party each obtained one seat. (In February 1998 two remaining seats were declared to have been won by the PPP/CIVIC alliance.).

mid-January 1998: In response to continued unrest among opposition supporters, the Herdmanston Accord was signed by Jagan and the PNC leader, Hoyte, providing for the organization of fresh elections within 36 months, the creation of a commission to recommend constitutional reforms, and an independent audit of the December 1997 election.

early June 1998: The CARICOM electoral team responsible for the independent audit of the election declared that there had been only minor procedural irregularities.

late June 1998: Owing to its boycott of the National Assembly, the PNC deputies forfeited their legislative seats, having failed to attend six consecutive sittings. Subsequently, legislation designed to enable the PNC deputies to recover their seats in the legislature was formulated by both sides, and the PNC (with the exception of Hoyte, who continued to deny the legitimacy of Jagan's authority) rejoined the National Assembly on 14 July.

January 1999: A 20-member Constitutional Reform Commission was created, comprising representatives of the country's principal political parties and community groups, to formulate recommendations for constitutional reform. The Commission published a report of its proposals in July.

March 1999: Guyana and Venezuela established the High Level Binational Commission, intended to expedite the resolution of the territorial dispute and to promote mutual co-operation.

late April 1999: The Guyana Public Service Union (GPSU) arranged a general strike after the Government had offered an unsatisfactory pay rise. Other public-sector unions later joined the GPSU in taking industrial action, and much of the country was paralysed. After 56 days of industrial action, an interim agreement was reached, pending a full settlement of their claim by an arbitration tribunal. Following the decision of the tribunal, however, the Government announced that it would only be bound by the tribunal's findings if it could be proved that the tribunal had followed the correct procedures, and warned the unions that the cost of the pay rises might necessitate redundancies.

11 August 1999: President Jagan announced her retirement on the grounds of ill health, and was replaced by Bharrat Jagdeo, who reappointed Hinds as Prime Minister (Hinds had earlier resigned in accordance with the requirements of the Constitution).

October 1999: President Hugo Chávez Frías of Venezuela announced his Government's intention to reopen its territorial claim.

May 2000: Suriname formally claimed that Guyana had violated its territorial integrity by granting a concession, to a Canadian-based company, CGX Energy Inc, to explore for petroleum and gas in a disputed maritime area. In June the two countries initiated negotiations to settle the dispute.

July 2000: The National Assembly approved legislation establishing a permanent electoral-reform task force, the Guyana Elections Commission (GECOM).

October 2000: The National Assembly approved a constitutional amendment establishing a mixed system of proportional representation, combining regional constituencies and national candidate lists; it also approved the abolition of the Supreme Congress of the People of Guyana and the National Congress of Local Democratic Organs.

17 November 2000: Delays in the enactment of the electoral reform resulted in the postponement of elections to 19 March 2001. The Government declared its intention to continue to govern until March, while the opposition insisted that the Government should resign and that an interim coalition Government should be formed. It was subsequently decided that the PPP/CIVIC would remain in office, but would limit its powers.

15 January 2001: A constitutional crisis arose when Justice Claudette Singh declared the December 1997 elections to be null and void: she ruled that the legislature had acted illegally in making the possession of a voter identification card a prerequisite for voting, and alleged that there had been instances of electoral fraud. However, she later ruled that the Government should remain in office until the March ballot, although no legislation, other than that necessary for the election, should be enacted during this time.

19 March 2001: In the general and regional elections the PPP/CIVIC gained 53.0% of the votes cast, obtaining a majority in the National Assembly, with 34 seats; the PNC (contesting the elections as the PNC/Reform) won 27 seats, the Guyana Action Party, in alliance with the WPA, gained two seats, while Rise, Organize and Rebuild Guyana Movement and TUF each secured one seat. An injunction applied for by the PNC/Reform, alleging breaches of electoral procedure, was dismissed by Chief Justice Desirée Bernard on 31 March. Jagdeo was declared President-elect.

May 2001: The Government announced the depoliticization of the civil service, with a new head of the service being created, outside the Office of the President.

Haiti

AD 650: The island which came to be known as Hispaniola (or Quisqueya, the Amerindian name), and which was later divided into the Dominican Republic and Haiti, was inhabited by Taino.

1492: The Italian navigator in the service of the Spanish Crown, Cristoforo Colombo (Christopher Columbus), landed on the island and named it Hispaniola ('Little Spain'). Columbus founded a small settlement, Navidad, on Hispaniola.

1493: The first permanent European settlement, named Isabella, was founded on the island.

1520s: The Spanish imported African slaves to replace the Amerindian labourers who had disappeared as a result of exploitation and illness.

1697: The western part of the island was ceded to France as Saint-Domingue by the Treaty of Ryswick. The remaining section, controlled by the Spanish, was called Santo Domingo (now known as the Dominican Republic).

1795: France gained nominal control of the whole island through the Treaty of Basle.

1 January 1804: An uprising by African-descended slaves in 1794, led by Toussaint Louverture, resulted in the establishment of Haiti as an independent state. Jean-Jacques Dessalines proclaimed himself Emperor.

1807–18: Following the assassination of Dessalines in 1806, Alexandre Pétion ruled the country.

1822: The President of Haiti, Jean-Pierre Boyer, annexed Santo Domingo, and abolished slavery.

1844: Haitian rule was overthrown through revolution, and Santo Domingo declared its independence, forming the Dominican Republic.

1915–34: The USA, fearing European economic and political rivalry in its Caribbean sphere of influence, invaded Haiti, ostensibly to restore order.

1946: A military *coup d'état* overthrew President Elie Lescot's Government and the liberal Dumarsais Estimé was appointed President.

October 1950: Gen. Paul Magloire was elected to the presidency for a six-year term.

12 December 1956: Magloire resigned following widespread public disorder and a national strike. The President of the Supreme Court, Pierre-Louis, was subsequently inaugurated as Provisional President.

22 September 1957: Dr François ('Papa Doc') Duvalier was elected President.

14 June 1961: Duvalier declared himself re-elected for another six-year term.

17 November 1961: Several foreign diplomats claimed that members of a notorious private army loyal to President Duvalier, popularly known as the Tontons Macoutes (Creole for 'Bogeymen'), had attempted to extort money from them. The Haitian Government requested that the British Government recall its Ambassador.

22 April 1964: Duvalier was elected President-for-life in a referendum in which all the ballot papers were officially reported to have approved the proposal.

1968–70: Several attempts to overthrow the Duvalier regime were violently repressed by the security forces and the Tontons Macoutes.

12 February 1971: A referendum approved a reduction in age restrictions on the presidency, that would have debarred Duvalier's son from succeeding his father.

21 April 1971: Duvalier died, and was replaced as President by his son, Jean-Claude Duvalier.

11 February 1973: The first elections to the National Assembly since 1961 were held. The candidates, all of whom supported the President, campaigned on local issues. No political parties were allowed.

11 February 1979: The Parti de l'Unité Nationale, Haiti's only legal party, won 57 of the 58 seats available in the National Assembly amid widespread allegations of electoral fraud.

10 April 1983: Voting took place in the first round of municipal elections. The second round took place on 15 May, but voting was postponed indefinitely in the capital, Port-au-Prince. All candidates required prior government approval.

27 August 1983: The National Assembly unanimously approved a new Constitution, reaffirming the younger Duvalier's status as President-for-life and allowing him to appoint his successor.

12 February 1984: At legislative elections pro-Duvalier candidates won all 59 seats in the newly expanded National Assembly. Opposition candidates had previously been prevented from standing and serious electoral irregularities were reported.

22 July 1985: Reforms to the Constitution proposed by President Duvalier were approved at a referendum. The proposals provided for the legalization of political parties by 1987 and for the establishment of the post of Prime Minister, who was to be appointed by the President. Only pro-Duvalier parties would, however, be approved.

31 January 1986: President Duvalier announced a 30-day 'state of siege' and imposed martial law in the wake of large-scale violent demonstrations against his Government.

7 February 1986: President Duvalier fled the country, with assistance from the US Government. Gen. Henri Namphy, the army Chief of Staff, announced that a five-member Conseil National Gouvernemental (CNG) would form an interim administration. Namphy later formed a Cabinet, that included a majority of civilians.

19 October 1986: Elections to a Constituent Assembly, that was to propose reforms to the existing Constitution, were widely boycotted by voters and by opposition parties, which called on the CNG to resign.

29 March 1987: A referendum overwhelmingly approved the Constitution proposed by the Constituent Assembly.

29–30 November 1987: The authorities cancelled scheduled general elections after several violent incidents during polling left at least 34 people dead. In an attempt to appease rioters, Namphy subsequently announced the dissolution of the CNG.

17 January 1988: Presidential and legislative elections were held, despite numerous accusations of electoral malpractice. Prof. Leslie Manigat, the leader of the Rassemblement des Démocrates Nationalistes et Progressistes (RDNP), was elected President and appointed a Government largely composed of independents.

20 June 1988: Namphy led a military *coup d'état*, which overthrew Manigat's Government. Namphy, who named himself President, subsequently abrogated the Constitution.

18 September 1988: Members of the Presidential Guard overthrew Namphy's Government and installed Brig.-Gen. Prosper Avril as President. The soldiers rebelled after Namphy announced his intention to arrest several of their colleagues for disloyalty.

10 March 1990: Avril resigned as President after popular opposition to his regime increased, and was succeeded by a Supreme Court judge, Ertha Pascal-Trouillot.

15 August 1990: The Council of State formally censured Pascal-Trouillot on the grounds of financial incompetence, corruption, nepotism and failure to impose law and order. The President announced that she would hold elections later that year.

16 December 1990: Fr Jean-Bertrand Aristide, of the Front National pour le Changement et la Démocratie (FNCD), won a convincing victory at presidential elections. Aristide became Haiti's first-ever democratically elected President.

30 September 1991: Aristide was deposed in a military *coup d'état* led by the Commander-in-Chief of the armed forces, Brig.-Gen. Raoul Cédras. The rebels, discontented with the President's attempts to reduce military influence, forced Aristide into exile. The *coup d'état* was condemned by the Organization of American States (OAS), which imposed economic sanctions on Haiti, and continued to recognize Aristide as the legitimate President.

7 October 1991: Soldiers coerced the National Assembly into approving the appointment of a Supreme Court judge, Joseph Nerette, as interim President.

February 1992: Following negotiations an agreement was reached, providing for the installation of the leader of the Mouvement pour la Reconstruction Nationale, René Théodore, as Prime Minister; Théodore was to govern in close consultation with Aristide.

March 1992: Following an appeal by Nerette, the Supreme Court annulled the agreement reached in February, declaring it unconstitutional.

May 1992: Following negotiations between the legislature, the Government and the armed forces, an agreement providing for a 'consensus government' was ratified by the Senate. Provisions of the agreement included the appointment of a new Prime Minister and a multi-party government.

19 June 1992: Nerette resigned as President and was not replaced. Marc Bazin, of the Mouvement pour l'Instauration de la Démocratie en Haïti (MIDH), and who had finished second to Aristide in the 1990 presidential elections, was appointed Prime Minister. Bazin called on Aristide to return to Haiti and resume the Presidency.

3 July 1993: Cédras and Aristide signed a 10-point peace agreement brokered by the UN and the OAS.

30 August 1993: Aristide appointed Robert Malval as Prime Minister under the terms of the peace agreement.

14 October 1993: Extremists opposed to the return of Aristide assassinated the Minister of Justice, Guy Malary, and the Malval Government was obliged to operate in a state of virtual siege.

17 December 1993: Malval resigned as Prime Minister on the grounds that he was unable to ensure the successful return of Aristide to Haiti.

11 May 1994: The Cédras regime announced the creation of a civilian Government headed by the President of the Court of Cassation, Emile Jonassaint, who took the title of Provisional President. The USA, however, refused to recognize the new administration.

23 May 1994: The UN introduced more stringent economic sanctions in an attempt to force Cédras to resign and allow Aristide to return.

31 July 1994: UN Security Council Resolution 941 authorized the use of military force to overthrow the military regime.

19 September 1994: US troops intervened in Haiti in support of an agreement between Cédras and US diplomats, intended to promote a transition to democratic Government.

10 October 1994: Cédras resigned as leader of the military junta before leaving for exile in Panama. Jonassaint resigned his position as Provisional President three days later.

15 October 1994: Aristide returned to Haiti under US protection and was reinstated as President. He appointed Smarck Michel as his Prime Minister and pledged to hold early elections.

31 March 1995: The UN Mission in Haiti (UNMIH) was formally inaugurated and replaced the US troops stationed in the country.

25 June 1995: The first round of voting in legislative elections was held amid violence and accusations by the opposition of serious irregularities.

17 September 1995: A second round of voting was held. The electoral commission subsequently announced that the Plateforme Politiqe Lavalas (PPL) coalition, loyal to President Aristide, had won the majority of seats in both the Chamber of Deputies (Chambre des Députés) and the Senate (Sénat).

16 October 1995: Michel resigned following a disagreement with President Aristide over his support for privatization. The Minister of Foreign Affairs, Claudette Werleigh, replaced him as Prime Minister.

17 December 1995: René Préval, Aristide's chosen candidate, was elected President with 87.9% of the vote. Aristide had been constitutionally barred from seeking re-election.

5 January 1996: Préval successfully requested that the UN extend UNMIH's mandate for a further six months, from 1 March.

27 February 1996: The legislature ratified President Préval's appointment of Rosny Smarth as Prime Minister.

28 June 1996: UNMIH's mandate was extended until 30 November.

3 November 1996: Aristide founded a new political organization, La Fanmi Lavalas (FL), to promote his presidential candidacy for 2000.

27 March 1997: A vote of 'no confidence', arising from the Government's controversial implementation of a structural adjustment programme designed by the International Monetary Fund (IMF), was defeated in the legislature by 37 votes to 29.

6 April 1997: Elections were held for several seats in the Senate and Chamber of Deputies as well as for hundreds of seats on local councils. The authorities announced, however, that the elections would be rerun after no clear winners emerged in several constituencies.

9 June 1997: Following a month of violent demonstrations and industrial action held to protest at the Government's financial-austerity measures, Smarth resigned as Prime Minister.

26 August 1997: The Chamber of Deputies overwhelmingly rejected the appointment of Eric Pierre as Prime Minister and Smarth remained in office in an interim capacity.

3 November 1997: President Préval nominated Hervé Denis as the new Prime Minister, but the Chamber of Deputies again rejected the appointment, reportedly in protest at the failure to hold reruns of the April elections.

30 November 1997: UNMIH's mandate formally expired and UN troops left the country. Several hundred US soldiers remained in the country to guarantee internal security.

23 March 1998: President Préval nominated Denis as Prime Minister once again, but the appointment was again blocked by the legislature.

18 December 1998: Congress approved the appointment of the incumbent Minister of National Education, Jacques Edouard Alexis, as Prime Minister.

11 June 1999: The newly-appointed Conseil Electoral Provisoire (CEP) announced it was to disregard the results of the controversial April 1997 elections and would organize new elections.

6 October 1999: The CEP ruled that the legislative and municipal elections scheduled for November and December should be postponed.

3 April 2000: The CEP again announced the postponement of the legislative and municipal elections until mid-May.

21 May 2000: The first round of the legislative and municipal elections finally took place, amid widespread allegations of electoral fraud. The following month the CEP announced that Aristide's FL had won a majority in the legislature, but the opposition refused to recognize the outcome after the CEP President, Leo Manus, went into exile rather than approve the results. The results were criticized as inaccurate by the UN and the OAS.

9 July 2000: The second round of legislative and municipal elections was boycotted by international observers and by the opposition coalition, the Convergence Démocratique (CD), in protest at alleged electoral fraud at the first round. According to official results, the FL won 72 seats in the Chamber of Deputies and 18 of the 19 seats contested in the Senate.

20 September 2000: Following President Préval's announcement that presidential elections would be held in late November, the CD declared its intention to boycott them, unless the legislative and municipal elections were rerun.

26 November 2000: Aristide won 91.7% of the votes at the presidential elections, which were boycotted by the major opposition parties.

January 2001: The leaders of the CD announced the formation of a rival, provisional Government, the Front Alternatif, which pledged to organize new elections.

7 February 2001: Aristide was inaugurated as President. The opposition simultaneously designated Gérard Gourgue the leader of the Front Alternatif.

March 2001: President Aristide appointed Jean-Marie Chérestal as Prime Minister and it was announced that new legislative elections would be held in November 2002. The opposition rejected this offer and violent demonstrations took place in Port-au-Prince. Later that month the Senate passed a resolution calling for Gourgue's arrest.

May 2001: The former ruler, Gen. Avril, was arrested on charges of torturing opponents during his rule.

July 2001: Following mediation by the OAS, the Government announced that legislative elections would be held in 2002, although no timetable was agreed upon.

Honduras

5th century AD: Mayan civilization reached western Honduras and spread rapidly, with Mayan control centred on the major ceremonial area of Copán, near the present-day town of Santa Rosa de Copán.

17 August 1502: The navigator and explorer Cristoforo Colombo (Christopher Columbus) landed near what is now the town of Trujillo in Honduras.

1524: The invasion of Honduras by four separate Spanish expeditions marked the beginning of a series of conflicts among the Spaniards and with the indigenous population.

1539: Spain's conquest of Honduras was completed.

15 September 1821: Honduras gained independence from Spain, in a federation with the other four Central American states, Costa Rica, El Salvador, Guatemala and Nicaragua.

1823: The United Provinces of Central America was formed, under the rule of Gen. Manuel José Arce.

1838: Following the dissolution of the United Provinces of Central America, Honduras became a sovereign state, formally declaring its independence on 15 November.

January 1839: Honduras adopted an independent Constitution.

1841: Gen. Ferrera, a Conservative, was elected as Honduras' first post-independence President and effectively controlled the country until 1847.

1848: A new Constitution was promulgated, under the presidency of another Conservative, Juan Lindo Zelaya.

1852–55: The Liberals, led by Trinidad Cabañas, held power in Honduras.

1855: Cabañas was replaced by Santos Guardiola, a Conservative, following the invasion of Honduras by Guatemalan forces.

1859: The British Government agreed to recognize Honduran sovereignty over the Islas de la Bahía.

1862: Guardiola was assassinated by his honour guard; the presidency changed hands frequently in the next 14 years.

late 19th century: The export of Honduran bananas to the USA commenced; large areas of land were subsequently granted to US fruit companies, often in exchange for political support.

1876: A Liberal, Marco Aurelio Soto, became President of Honduras, with the support of President Justo Rufino Barrios of Guatemala.

1883: Following further intervention from Barrios, Soto was forced to resign and was replaced by Gen. Luis Bográn Baraona.

1891: Gen. Poinciana Leiva, who had ruled briefly in the mid-1870s, was returned to the presidency.

1894: Following a period of civil conflict, Policarpo Bonilla, the leader of the nascent Partido Liberal (PL), assumed power.

1895: A new Constitution was adopted, and Policarpo Bonilla was formally elected to the presidency.

1899: Policarpo Bonilla was succeeded by his military commander, Gen. Terencio Sierra, at the former's instigation.

1903: Sierra was deposed by Gen. Manuel Bonilla, a Conservative.

1906: A commission that had been established in 1894 to resolve a protracted border dispute between Honduras and Nicaragua allocated the majority of the disputed territory to Honduras.

1906: Honduras signed friendship agreements with El Salvador and Guatemala.

1907: The Presidents of Costa Rica, El Salvador, Guatemala, Honduras and Nicaragua signed a Treaty of Peace and Friendship, in which they pledged themselves to establish a Central American Court of Justice to resolve future disputes; a special US-sponsored clause provided for Honduran neutrality in any future Central American conflict.

February 1907: Exiled Honduran Liberals, supported by elements of the Nicaraguan army, invaded Honduras and succeeded in overthrowing Bonilla; the USA, apparently concerned about Nicaragua's potential dominance in regional affairs, dispatched troops to the Honduran town of Puerto Cortés to protect the banana trade; Gen. Miguel Dávila, a Liberal, was installed as President.

1908: An uprising led by opponents of President Dávila, and allegedly supported by El Salvador and Guatemala, ultimately failed.

1911: Following a further rebellion, headed by former President Manuel Bonilla, and mediation by the USA, Dávila resigned and was replaced by an interim President, Francisco Bertrand.

1912: Manuel Bonilla was elected President.

1913: Following Bonilla's death, Bertrand returned to the presidency and remained in power until 1920.

1917–20: Workers in the banana industry took industrial action, although the banana companies, which held considerable power in Honduras, continued to prosper.

1920–23: Some 17 attempted *coups d'état* or rebellions were instigated in Honduras.

October 1920: Gen. López Gutiérrez was inaugurated as President of Honduras.

1923: A second Treaty of Peace and Friendship was adopted by Costa Rica, El Salvador, Guatemala, Honduras and Nicaragua.

October 1923: No candidate secured a majority of votes in presidential elections, and the legislature, which, under the terms of the Constitution, was to elect the President in such an event, proved unable to reach a decision.

1924: A new Constitution was adopted.

January 1924: After López Gutiérrez refused to schedule new elections, Gen. Tiburcio Carías Andino declared himself President, provoking armed conflict.

April 1924: A peace agreement was signed, providing for an interim presidency, held by Gen. Vicente Tosta.

28 December 1924: Miguel Paz Barahona, the candidate of the Partido Nacional (PN), was elected President of Honduras, following Carías' decision to withdraw his candidacy and the PL's refusal to nominate a candidate.

1928: The PL's Vicente Mejía Colindres defeated Gen. Carías, the PN candidate, in elections to the presidency.

16 November 1932: Carías was inaugurated as President of Honduras.

1936: A Constituent Assembly, appointed by President Carías, drafted a new Constitution, which specified that the incumbent President would remain in office until 1943; subsequent attempts by the PL and other opponents of the Government to depose Carías were all unsuccessful.

1939: The legislature extended, until 1949, the presidential term of Carías, who proceeded to rule Honduras as a dictatorship.

10 October 1948: In presidential elections Juan Manuel Gálvez, of the PN, was elected unopposed.

1951: The Organization of Central American States was founded by Costa Rica, El Salvador, Guatemala, Honduras, Nicaragua and Panama.

10 October 1954: Dr José Ramón Villeda Morales, the leader of the PL, won most votes in elections to the presidency, but lacked a majority, prompting a political crisis, as the main parties failed to agree on how to proceed; Julio Lozano Díaz, the incumbent Vice-President, subsequently suspended the legislature and declared himself President, Gálvez having left the country.

21 October 1956: Lozano Díaz was overthrown by a military junta.

October 1957: The PL secured a majority in Congress.

November 1957: Villeda was elected as President of Honduras.

8 January 1959: The Central American Common Market, comprising El Salvador, Guatemala, Honduras, Nicaragua, and later Costa Rica, came into force.

3 October 1963: Villeda was overthrown in a military uprising led by Col (later Gen.) Oswaldo López Arellano, the Minister of Defence.

June 1965: López was appointed President, following elections held on the basis of a new Constitution.

June 1969: Ill-feeling concerning the treatment of football supporters during three World Cup qualifying round matches between the El Salvador and Honduras national teams resulted in the countries suspending diplomatic relations. About 12,000 Salvadorean migrants resident in Honduras were expelled.

3–18 July 1969: Military skirmishes between Salvadorean and Honduran armed forces culminated, on 14 July, in Salvadorean troops advancing some distance into Honduras. After the intervention of the Organization of American States (OAS), a cease-fire was agreed on 18 July. A final peace treaty was not signed until 1980, and the underlying territorial dispute, which was referred to the International Court of Justice (ICJ) in 1986, was not resolved until 1992. About 2,000 people were killed in the conflict, which is sometimes known as the 'Football War'.

28 March 1971: The PN narrowly won presidential and congressional elections, and Dr Ramón Ernesto Cruz Uclés became President.

4 December 1972: Cruz was deposed in a bloodless *coup d'état*, and former President López was reinstalled as head of state.

September 1974: Some 10,000 people were killed as a result of Hurricane Fifi, which also severely damaged the Honduran economy.

31 March 1975: At the instigation of the military, López was replaced as Commander-in-Chief of the Armed Forces by Col (later Gen.) Juan Alberto Melgar Castro.

22 April 1975: Melgar was appointed as President.

7 August 1978: Melgar was forced to resign and was replaced by a military junta, led by the Commander-in-Chief of the Armed Forces, Gen. Policarpo Paz García, as head of state.

early 1980s: Former members of the Nicaraguan National Guard, regarded by the Sandinista Government of Nicaragua as counter-revolutionaries ('Contras'), began to establish bases in Honduras, from which they conducted raids across the border between the two countries, allegedly with support from the Honduran armed forces.

20 April 1980: Elections to a Constituent Assembly were held, marking an end to military rule; the PL won 52% of the votes cast, and Gen. Paz was subsequently appointed as interim President, pending elections.

30 October 1980: The Governments of El Salvador and Honduras signed a peace treaty.

29 November 1981: The PL gained an absolute majority in elections to the National Assembly, while its leader, Dr Roberto Suazo Córdova, was elected to the presidency.

27 January 1982: Suazo was sworn in as President, although Col (later Gen.) Gustavo Alvarez Martínez, who was appointed head of the armed forces, held considerable power.

November 1982: The National Assembly approved a new Constitution, which provided for a democratic, republican system of government, with executive power held by the President (elected for a single four-year term) and legislative power vested in the 128-member National Assembly (elected by universal adult suffrage for a term of four years). Gen. Alvarez became Commander-in-Chief of the Armed Forces, following the constitutional separation of this post from that of President.

February 1983: The USA and Honduras initiated a series of joint military manoeuvres on Honduran territory, which continued throughout the 1980s, enabling the USA to construct permanent military installations in Honduras.

31 March 1984: Gen. Alvarez was deposed as Commander-in-Chief of the Armed Forces by a group of army officers.

24 November 1985: In accordance with new legislation providing for the winner of presidential elections to be the leading candidate from the party receiving the largest number of combined votes, José Simeón Azcona del Hoyo, the leading candidate of the PL, was elected President, although Rafael Leonardo Callejas Romero, of the PN, had individually secured by far the highest number of votes cast. The PL won 67 seats in elections to an enlarged 134-member National Assembly.

1 October 1986: The conflicting territorial claims of Honduras and El Salvador over three islands in the Gulf of Fonseca and a small area of land on the common border were submitted to the International Court of Justice (ICJ) for arbitration.

December 1986: President Azcona requested the departure of the Nicaraguan Contras from Honduras.

7 August 1987: Costa Rica, El Salvador, Guatemala, Honduras and Nicaragua signed a Central American peace agreement, the provisions of which included the implementation of simultaneous cease-fires in El Salvador and Nicaragua, a halt to foreign assistance to rebel groups and the establishment of national reconciliation commissions in each of the Central American nations.

1988 and 1989: The Inter-American Court of Human Rights found the Honduran Government guilty of the 'disappearances' of Honduran citizens between 1981 and 1984.

March 1988: Several thousand US troops were temporarily deployed in Honduras, in response to an incursion into Honduran territory by the Nicaraguan army; further violations of the border subsequently occurred, as Nicaraguan troops forced at least 12,000 Contra rebels into Honduras.

13–14 February 1989: At a summit meeting of the five Central American Presidents, it was agreed that the Nicaraguan Contra forces encamped in Honduras would demobilize; the conditions for this demobilization were expanded at a further summit meeting in August.

7 November 1989: The UN Security Council authorized the creation of the UN Observer Group for Central America (ONUCA), a multinational military force, to monitor developments in the region.

26 November 1989: The PL gained a majority of seats in elections to the National Assembly, while the PN's Callejas won the concurrent presidential election.

June 1990: The Contra units officially disbanded and left Honduras.

11 September 1992: The ICJ awarded Honduras sovereignty over some two-thirds of the mainland territory it disputed with El Salvador and over one of the three disputed islands in the Gulf of Fonseca.

16 March 1993: The Government established a special commission to investigate persistent allegations of human rights violations by the armed forces.

28 November 1993: The PL's Carlos Roberto Reina Idiaquez, a former President of the Inter-American Court of Human Rights, was elected President, winning 52% of the votes cast, while his party also secured a clear majority in the National Assembly, which had been reduced to 128 members.

4 May 1994: The National Assembly approved a proposal for a constitutional amendment to abolish compulsory military service and the transfer of the police from military to civilian control, although conscription was temporarily reintroduced in August of that year.

July 1994: Some 4,000 members of indigenous organizations occupied the National Assembly building for three days and succeeded in securing from the Government rights and social assistance for Honduras' indigenous population.

August 1994: The Government declared a state of national emergency in response to increasing public unrest resulting from the effects of austerity measures, an energy crisis and food shortages.

23 January 1995: A new civilian criminal investigation body, the Dirección de Investigación Criminal, replaced the armed forces' much-criticized secret counter-intelligence organization, the División Nacional de Investigaciones.

June 1995: Honduras and Nicaragua signed an accord providing for the visible demarcation of each country's territorial waters in the Gulf of Fonseca, following frequent disputes regarding fishing rights; conflict continued, however, and the demarcation process did not begin until May 1998.

July 1995: Protests by indigenous organizations that the Government had not honoured its commitments of July 1994 resulted in renewed government pledges.

15 September 1995: The PL, the PN, the Partido Innovación y Unidad—Social Democracia (PINU) and the Partido Demócrata Cristiano de Honduras (PDCH) established the Consejo Nacional de Convergencia in an attempt to seek a consensus on political, social and economic issues.

30 November 1997: Carlos Roberto Flores Facussé, the candidate of the ruling PL, was elected President of Honduras, winning 52.7% of the votes cast. In concurrent elections to the National Assembly the PL also obtained a majority, winning 67 seats, while the PN secured 55 seats, the PINU three, the PDCH two and the left-wing Partido de Unificación Democrática the remaining seat.

19 January 1998: The Presidents of Honduras and El Salvador signed a convention specifying the rights and nationality of those people resident in territory reallocated following the ICJ ruling of September 1992.

May 1998: The police force was officially transferred from military to civilian control.

November 1998: President Flores declared a state of emergency in response to the devastation caused by Hurricane Mitch, which had struck Honduras in late October, killing at least 6,600 people and causing damage estimated at US $5,000m.

27 January 1999: The post of Commander-in-Chief of the Armed Forces and its responsibilities were transferred to the Minister of National Defence and Public Security.

July 1999: Flores implemented a number of changes to the military high command, following intelligence reports of a plot by senior-ranking officers to overthrow the Government, which was believed to have been prompted by resentment over civilian control of the armed forces.

1 December 1999: Nicaragua imposed a 35% tariff on imports from Honduras, following the latter's ratification of a maritime border treaty with Colombia, to which Nicaragua objected.

7 February 2000: Honduras and Nicaragua agreed to establish a maritime exclusion zone, following OAS mediation.

19–25 February 2000: Honduran and Nicaraguan patrol boats exchanged gunfire on three occasions in the Gulf of Fonseca.

7 March 2000: Honduras and Nicaragua signed an accord on joint patrols in the Caribbean, pending a ruling by the ICJ, and on combined operations in the Gulf of Fonseca, as well as on the withdrawal of forces from the land border area.

February 2001: The Nicaraguan defence minister accused Honduras of violating the March 2000 accord by carrying out military exercises in the area.

12 March 2001: The National Elections Tribunal finally accepted the presidential candidacy of the PN's Ricardo Maduro, following a dispute over his nationality; Rafael Pineda Ponce of the PL was among a number of candidates already registered for elections due to be held on 25 November.

May 2001: The National Assembly approved a change in electoral legislation that would allow Honduran nationals living in the USA to vote in the forthcoming general election.

7 June 2001: The foreign ministers of Honduras and Nicaragua signed a confidence-building agreement, following OAS mediation.

25 July 2001: Honduras declared a state of emergency in eight provinces where drought had resulted in severe food shortages.

Jamaica

AD 650: The Arawak Taino people settled on the island, which they named Xaymaca, or Isle of Springs.

1494: The Italian navigator in the service of the Spanish Crown, Cristoforo Colombo (Christopher Columbus), sighted the island.

1509: The island became a Spanish colony.

c. **1523:** The St Iago de la Vega settlement (later Spanish Town) was founded.

1655: A British naval force led by Sir William Penn captured the island.

1670: The island formally came under British control, through the Treaty of Madrid. African slaves were imported as labourers, following the disappearance of the native Arawaks as a result of diseases brought by European settlers.

1 August 1838: Slavery was abolished.

October 1865: A rebellion in Morant Bay over economic and social issues, was suppressed, resulting in the death of some 500 people. The rebellion led to the abolition of the island's Assembly and the imposition of Crown Colony control by the United Kingdom.

1884: Representative government was partially restored.

1938: The centre-left People's National Party (PNP) was established by Norman Manley.

1943: The right-wing, anti-Communist Jamaica Labour Party (JLP) was established.

1944: Jamaica was granted a new Constitution.

24 May 1947: The Governor, Sir John Huggins, banned all public demonstrations for 28 days after the PNP organized a series of protests against the Government's handling of the island's high unemployment rate. The PNP none the less continued to hold rallies.

20 December 1949: The JLP retained power at legislative elections, winning 17 of the 32 seats.

12 January 1955: The PNP won a majority at legislative elections, and Manley formed a new Government.

2 August 1956: The British Parliament adopted a bill establishing the British Caribbean Federation, which was to have a joint Government, Legislature and Supreme Court.

14 March 1957: The Governor of Jamaica, Sir Hugh Foot, announced proposals to create virtual self-government on the island and a ministerial system of government. The proposals were approved by the Jamaican legislature.

4 July 1959: The new Constitution came into force by which Jamaica received full internal self-government within the West Indies Federation. The United Kingdom retained responsibility for external affairs and defence.

28 July 1959: The PNP were victorious at the first general elections held under the new Constitution and Manley was re-elected Prime Minister.

26 August 1959: The opposition JLP boycotted the opening of Congress in protest at alleged electoral irregularities.

19 September 1961: In a referendum some 54% of Jamaicans voted to withdraw from the West Indies Federation.

1–9 February 1962: The Jamaica Independence Conference was held in London. Government delegates and opposition members reached an agreement on the island's full independence with representatives of the British Government.

10 April 1962: The JLP defeated the PNP at general elections, and Sir Alexander Bustamante became Premier.

31 May 1962: The British Government formally dissolved the moribund West Indies Federation.

6 August 1962: Jamaica gained independence from the United Kingdom.

21 February 1967: The JLP increased its majority at legislative elections held in an atmosphere of violence and intimidation. Donald Sangster was appointed Prime Minister after Bustamante resigned on the grounds of ill health.

11 April 1967: Sangster died and was replaced as Prime Minister by the former Minister of External Affairs, Hugh Shearer.

29 February 1972: At legislative elections the JLP was heavily defeated by the PNP, whose leader, Michael Manley, was sworn in as Prime Minister.

19 June 1976–5 June 1977: In response to escalating street violence and crime, the Government imposed a state of emergency and intensified operations by the security forces.

15 December 1976: The PNP retained its majority at a general election held after a campaign marred by serious violence.

8 March 1977: The PNP recorded an overwhelming victory in local council elections, winning 237 of the 269 seats available.

17 December 1979: Manley effected a major reorganization of his administration, primarily in an attempt to reduce Jamaica's large budgetary deficit.

30 October 1980: At legislative elections the PNP was defeated by the JLP, whose leader, Edward Seaga, formed a new Government. The PNP's popularity had suffered as a consequence of its decision to reject further funding from the International Monetary Fund (IMF), despite the country's weak economic position. More than 100 people were killed in political violence prior to the election.

29 October 1981: Seaga announced that Jamaica was to sever diplomatic relations with Cuba, after the Cuban Government failed to extradite three men to face murder trials in Jamaica.

26 November 1983: Seaga announced an unexpected general election following demands for his resignation by the opposition over his handling of a major devaluation of the Jamaican dollar.

15 December 1983: The JLP won all 60 seats at legislative elections after the PNP refused to contest the election, claiming the JLP was in breach of an agreement between the two parties stating that no further elections should be held until new voter lists had been compiled.

24–27 June 1986: A general strike was held in protest at Seaga's policy of public-sector pay cuts.

29 July 1986: The PNP gained control of 11 of the 13 parishes contested at local elections. Several people were killed during the campaigning and the Governor-General subsequently launched an inquiry into allegations of electoral malpractice. Seaga offered to resign in response to the JLP's poor performance, but was persuaded to remain in his position.

9 February 1989: The PNP secured a significant victory at a general election, winning 45 of the 60 seats available in the House of Representatives; Manley was appointed Prime Minister.

6 March 1990: At elections to the 13 municipal councils, the distribution of seats remained unchanged.

31 December 1991: The Deputy Prime Minister, Percival Patterson, was dismissed, following his implication in a financial scandal.

15 March 1992: Manley resigned as Prime Minister for health reasons.

28 March 1992: A PNP Congress elected Patterson to replace Manley as Prime Minister. Patterson pledged to continue Manley's programme of economic liberalization.

30 March 1993: The PNP secured a resounding victory at legislative elections, and Patterson was reappointed Prime Minister. The elections were, however, marred by allegations of electoral malpractice and the JLP announced its intention to boycott subsequent by-elections, pending the introduction of electoral reform.

7 January 1995: Patterson effected a major reshuffle in an attempt to reduce the size of the Cabinet.

18 December 1997: At legislative elections the PNP won 56% of votes cast and 50 of the 60 seats in the House of Representatives. The JLP obtained 39% of the votes and secured 10 seats. Patterson was subsequently reappointed Prime Minister.

10 September 1998: Local elections, which had been repeatedly postponed since 1993 following allegations of electoral malpractice, finally took place. The PDP gained control of all 13 local councils. Seaga resigned from the JLP leadership for three months in protest at the party's decision to contest the elections.

13 July 1999: The Government deployed large numbers of troops throughout the capital, Kingston, in an attempt to combat the high incidence of violence and criminal activity, the majority of which was reportedly linked to drugs-trafficking.

8 March 2001: The PNP lost a seat to the JLP at a by-election in North-East St Ann.

11 March 2001: The leader of the opposition National Democratic Movement (NDM), Bruce Golding, resigned in response to the party's poor performance at the North-East St Ann by-election.

27 May 2001: Hyacinth Bennett was elected leader of the NDM.

June 2001: At least 27 people were killed in clashes in Kingston, between PNP and JLP supporters, and also with the security forces. Patterson deployed the Jamaica Defence Force to assist in quelling the violence in the capital.

Mexico

c. **10,000 BC:** Hunter-gatherer peoples migrated to the region from North-East Asia.

1500 BC–600 BC: The Olmec civilization existed in the Tabasco and Veracruz regions. They experimented with mathematics and developed a primitive calendar.

500 BC: Monte Alban was founded by the Zapotecs in the Oaxaca region. It was a highly organized state whose inhabitants developed an early form of pictorial written communication.

AD 100–650: Teotihuacán emerged as an important cultural and commercial city-state in the central area of present-day Mexico; its citizens constructed two large religious monuments, the Pyramid of the Sun and the Pyramid of the Moon.

250–900: The Mayan civilization occupied the areas of Chiapas, Tabasco, Campeche, Yucatán and Quintana Roo, as well as other parts of Central America. The civilization developed a highly complex system of hieroglyphics and is believed to have produced one of the first accurate calendars.

1100: A group of tribes, known as the Aztecs, of which one of the main tribes was the Mexica, migrated to the Valley of Mexico, where they encountered hostility from local inhabitants.

14th century: The Aztecs founded the settlement of Tenochtitlán on an island in the Lago de Texoco (near the site of present-day Mexico City).

1376: A monarchy was established, Acamapichtli being crowned as the first king of the Aztecs.

1519: The Spanish adventurer, Hernán (Fernando or Hernando) Cortés, led an expedition from the Spanish possession of Cuba, landing at what is now Veracruz. He formed an alliance with the Tlaxcalan people, who were dissatisfied at being subjugated by the Aztecs and, with his assistance, destroyed the important religious city of Cholula. They subsequently advanced to Tenochtitlán, where they captured the emperor, Moctezuma II, who died shortly afterwards.

1521: Following a siege of the city, during which many of its inhabitants died from an epidemic of smallpox and the last Aztec emperor, Cuauhtémoc, was captured, Tenochtitlán fell. The Spanish razed the ruined city and built the city of Mexico in its place.

early 16th century: Through an alliance of the city states of Tenochtitlán, Texoco and Tlacopan, the Aztecs had extended their influence throughout South America. A strict social hierarchy existed, headed by the monarchy and an appointed emperor. A number of deities were worshipped, including the sun and the serpent, Quetzalcoatl, the god of wind and learning.

1528: An *audiencia* was established by the Spanish Government, comprising four judges who were to impose Spanish rule in the colony. The newly-acquired lands were divided into estates called *encomiendas*, which were granted to soldiers and noblemen. The Europeans subsequently forced the indigenous people to work on their land.

1535: Antonio de Mendoza was appointed the first Viceroy of New Spain, a territory which was eventually to include the rest of Central America, the Antilles, the present-day US states of California and New Mexico, and the Philippines. He was assisted by the *audiencia*. The Spanish crown prevented the colonies from trading with any country other than Spain and decreed that one-fifth of colonial produce would be requisitioned by the monarch.

1542: The New Laws were introduced, which prohibited the enslavement of the indigenous peoples and made obsolete the practice of granting *encomiendas*.

1546–52: Following the earlier discovery of silver in the region, mines at Zacatecas, Real del Monte, Pachua and Guanajuato produced what became the colony's major export.

1551: The University of Mexico was established.

1571: Courts of the Spanish Inquisition were introduced in the colony to enforce the Catholic doctrine and prosecute religious heretics.

1573: A special tribunal was founded to which indigenous peoples could bring complaints against district governors.

mid-17th century: The diversion of supplies of mercury—essential to the production of silver—to Potosí, in present-day Bolivia, precipitated the decline in the industry.

1765–71: The Spanish Government introduced a number of administrative reforms, dividing the colony into 12 provinces, each ruled by an *intendente* who reported to a commandant-general in Mexico City who was in turn directly accountable to the Spanish monarchy. Legislation was also introduced providing for a more efficient collection of taxes; this caused unrest in the colony.

1778: The restrictions on trade with other countries by the colony were lifted by the Spanish Government.

19th century: *Mestizos*, individuals of mixed European and indigenous American descent, came to form the largest ethnic group in the population.

1804: The Spanish Government seized land and assets held by the Roman Catholic Church in the colony.

1808: A group of Spanish-born senior officials instigated a military *coup d'état* against the Viceroy, following his appointment of Creoles (individuals of Spanish descent born in Mexico) to prominent positions. Pedro Garibay was subsequently installed in the office.

16 September 1810: A group of Creole insurgents, led by a radical priest, Miguel Hidalgo y Costilla, rebelled against what they perceived as their subjugation by Spanish-born colonists and captured the cities of Zacatecas, San Luís Potosí and Valladolid. The rebel army advanced to Mexico City, where they fought supporters of the Spanish monarchy.

March 1811: The rebel army was defeated at Mandova, where Hidalgo was captured. He was later convicted of heresy by the Holy Office of Inquisition and executed by firing squad. José María Morelos Pavón subsequently assumed the leadership of the Creole insurgents.

June 1813: A rebel congress was convened at Chilpancingo, where Mexican independence was declared. The delegates drafted a document, which envisaged the introduction of universal male suffrage, popular sovereignty and the abolition of slavery in the new republic.

1815: Spanish forces captured Morelos and he too was executed; the independence movement disintegrated into a series of local guerrilla groups led by Guadalupe Victoria (Manuel Félix Fernández) in Puebla and Vicente Guerrero in Oaxaca.

1820: Moses Austin obtained permission from the Government for US citizens to settle the region of Texas (Tejas); settlement began the following year, led by his son, Stephen.

1821: Fearing unrest in the colony following a revolution in Spain, the leader of the royalist forces, Agustín de Iturbide, concluded an agreement with Guerrero, known as the Plan of Iguala, which guaranteed that Mexico would henceforth be an independent state, ruled as a limited monarchy, with the Roman Catholic Church as the state church. The viceroy was subsequently forced to resign and Iturbide proclaimed himself Emperor of Mexico.

March 1823: A group of senior army officers forced Iturbide to abdicate and proclaimed a republic; Guadalupe Victoria was appointed the first President. The other Central American territories in the former Spanish Empire seceded to form the United Provinces of Central America.

1824: A new Constitution was promulgated, which provided for the establishment of the Estados Unidos Mexicanos (United Mexican States), governed by a Senado (Senate) and a Cámara de Diputados (Chamber of Deputies). A President and Vice-President were to be elected for four-year terms, the President being able to exercise unlimited powers in states of emergency.

1827: The Vice-President led a conservative rebellion.

1828: A dispute arose concerning the results of a general election contested by Guerrero, as the liberal candidate, and the conservative representative, Manuel Gómez Pedraza. The opposing factions formed rival governments, the conservative administration eventually being forced to dissolve by the liberal forces.

1829: The Government enacted legislation abolishing the institution of slavery.

July 1829: A Spanish force, led by Gen. Isidro Barradas, attempted to reconquer Mexico; they were repelled by Mexican troops under the command of Gen. Antonio López de Santa Anna.

1830: Anastasio Bustamante led a revolt against Guerrero and installed himself as the third President.

1833: Following a counter-coup, Santa Anna was installed as President, although he later delegated the office to Valentín Gómez Farías.

April 1834: Santa Anna removed Gómez Farías and appointed himself President.

1836: The Texans rebelled in protest at amendments to the Constitution, which reduced the autonomy of the individual states. Their forces were first defeated by the Government at the Alamo, but gained a significant victory at the Battle of Jacinto, under the command of Sam Houston. In May, the Treaty of Velasco, between the Government and the Texans, stipulated that the Mexican troops were to retreat south of the Río Bravo (Río Grande) and a secret clause established the independence of the state, which was subsequently incorporated into the USA.

1844: Santa Anna was deposed and exiled to Cuba.

1845: Following the failure of the USA's attempts to negotiate the adjustment of the national borders with the Government of Mexico, US troops were deployed in the Río Bravo region.

1846: The advance of Mexican troops across the Río Bravo prompted the US Congress to declare war on the Government. Santa Anna returned to the presidency.

1847: The Maya of the Yucatán peninsula began a revolt against the white and mestizo population of the region in a conflict which became known as the Caste War of Yucatán.

September 1847: Following a series of heavy losses by government troops, Mexico City fell to the US forces; the Mexican army subsequently surrendered.

2 February 1848: The Mexican and US Governments signed the Treaty of Guadalupe Hidalgo, by which the Río Bravo was confirmed as the southern boundary of the state of Texas and the USA gained control of what became the state of New Mexico and parts of the states of Arizona, California, Colorado, Utah and Wyoming. Santa Anna resigned and went into exile.

1853: The Caste War ended in the defeat of the Maya, many of whom were forced to migrate to the state of Quintana Roo.

1853: Having returned from exile and installed himself as a dictatorial ruler, Santa Anna negotiated the Gadsden Purchase with the US Government, by which he sold the Mesilla Valley and southern Arizona to the USA for US $10m.

1854: Santa Anna became the subject of an anti-Government campaign by liberal groups, who forced him to return into exile; Juan Ruiz de Alvarez succeeded him as President.

1855: Alvarez was succeeded by Ignacio Comonfort, who proposed a series of reforms reducing the power of the church and the army.

1857: A new Constitution was promulgated, in which a federal system of government was reintroduced; universal suffrage and civil liberties, such as freedom of speech, were guaranteed, and the church was ordered to sell its land and buildings not in frequent use.

1858–60: A dispute between conservative and liberal factions in the Government over Comonfort's reforms resulted in a conflict known as the War of Reform. The liberal Government was forced to flee to Veracruz and the conservatives established a rival administration in Mexico City.

January 1861: Liberal forces recaptured Mexico City and Benito Juárez was elected President. Having already nationalized ecclesiastical property and enacted legislation separating the church and the state at Veracruz, Juárez announced that interest on payment on foreign loans made by previous administrations would be suspended.

October 1861: Angered at Juárez's policies, Spanish, British and French forces began an occupation of the eastern coast.

June 1863: The Spanish and British forces having withdrawn, the French forces of Napoleon III continued to make advances against the Mexican army. The French emperor invited the Austrian Archduke Ferdinand Maximilian Joseph von Habsburg to lead a new provisional government; however, Maximilian refused Napoleon's demands to revoke the Reform Laws.

1865: The US Government dispatched troops to assist Juárez against Napoleon.

19 February 1867: Following the return of Napoleon to France owing to developments in Europe, Juárez's forces compelled Maximilian to surrender; he was later executed.

December 1867: Juárez was elected President of a liberal Government. His attempts to consolidate the presidency by suspending some constitutional guarantees prompted liberal factions to accuse him of establishing a dictatorship.

1871: Following a presidential election in which neither Juárez nor his closest opponent, Gen. Porfirio Díaz, secured an outright majority, legislature declared Juárez President; Gen. Díaz subsequently launched an unsuccessful rebellion.

1872: Juárez died; he was succeeded by Sebastián Lerdo de Tejada.

1876: Gen. Díaz led a further insurrection; he succeeded in removing Lerdo and assumed the presidency himself. He subsequently enacted legislation directed against the exchange of contraband across the border with the USA, and honoured US citizens' claims worth US $4m. against Mexico.

1880: Díaz retired from the presidency, appointing Gen. Manuel González to succeed him. His administration supervised the development of an infrastructure and a telegraph network.

1883: The Government seized vast tracts of private and communal land in an attempt to attract foreign buyers. Large land companies instead purchased the estates, forcing the increasing population of landless peasants to labour for meagre wages.

1884: Díaz returned to the presidency; he resumed his policies of modernization and reform, persecuting and imprisoning his liberal critics.

1 July 1906: A group of exiled liberals proclaimed their intention to remove Díaz from office at a convention in St Louis, Missouri, USA. Their ideas were disseminated through the newspaper *Redención.*

1910: Díaz announced that he would accept a contestant at forthcoming presidential elections; however, when a liberal candidate, Francisco Indalécio Madero, proclaimed his candidacy, he was arrested. Díaz retained the presidency. Madero was released and fled to the USA, where he subsequently announced the Plan of San Luis Potosí, which demanded the restoration of democracy.

January 1911: A rebellion occurred in the state of Chihuahua, led by Pascual Orozco and Francisco 'Pancho' Villa.

November 1911: Madero was appointed President. When he failed to enact policies of reform, however, radicals revolted under the leadership of Emiliano Zapata. Zapata issued the Plan of Ayala, which demanded the immediate transfer of land to peasant farmers and the right of Mexican citizens to choose their own leaders; the proclamation fomented rebellion in the states of Morelos, Guerrero, Tlaxcala, Puebla and México.

February 1913: Félix Díaz, the nephew of Gen. Díaz, attempted a counter-coup. A general of Madero's Government, Victoriano Huerta, defected to the counter-revolutionaries' camp, seized control of Mexico City and appointed himself President. Madero was assassinated shortly afterwards.

March 1913: Venustiano Carranza led an army of Constitutionalists against Huerta. The US President, Woodrow Wilson, condemned the Huerta Government and dispatched troops to Veracruz, prompting anti-US protests.

8 July 1914: Huerta resigned. The revolutionary forces of Pancho Villa and Emiliano Zapata subsequently came into conflict with the Constitutionalists for control of the country.

December 1914: The revolutionary forces captured Mexico City and Puebla. They were subsequently compelled to retreat by supporters of Carranza, who had declared himself President.

August 1915: Following a campaign of terrorism by Mexican rebels in the border regions of the USA, a commission of eight Latin American countries and the USA recognized Carranza as President. Zapata consequently announced the end of his campaign.

March 1916: A number of raids into its territory by forces loyal to Pancho Villa prompted the US Government to dispatch a punitive expedition to the area, led by Gen. John J. Pershing.

December 1916: A constitutional congress met in Querétaro. The document promulgated the following year included a labour code, prohibited the President from serving consecutive terms and restricted the Roman Catholic Church's rights to land ownership.

May 1917: Carranza won presidential elections; his failure to enforce many of the provisions of the Constitution subsequently provoked opposition to his rule.

1918: Agitation by revolutionaries caused conflict in Morelos.

March 1919: Zapata wrote Carranza an open letter, in which he protested at the President's failure to enact sections of the Constitution; he was murdered the following month by Carranza's forces.

May 1920: A rebellion occurred, led by three of Carranza's generals: Plutarco Elias Calles, Alvaro Obregón and Adolfo de la Huerta. Carranza was killed in the conflict.

November 1920: Obregón was elected President.

1923: Following President Obregón's agreement to US petroleum companies' claims on deposits in Mexico, the US Government recognized the administration and later supported it during an abortive revolt. Calles subsequently succeeded Obregón as President.

July 1926: A series of religious reforms, which prohibited religious processions, closed religious schools and deported foreign nuns and monks, were condemned by the Archbishop of Mexico City, José Mora y del Río. A rebellion by Catholic insurgents ensued, during which thousands were killed.

1927: Dwight W. Morrow, the US Ambassador to Mexico, attempted to mediate between the Government and the Catholic rebels.

July 1928: Obregón was assassinated. Calles established the Partido Nacional Revolucionario (PNR), with himself as leader. He appointed a series of 'puppet' presidents from among its members, who instituted his rule in the ensuing years.

1934: Lázaro Cárdenas won presidential elections. He sought to reduce the role of the army in government affairs and introduced a number of reforms in the areas of social welfare, land and education. Following his criticism of these policies, Calles was forced into exile.

1936: The Government enacted legislation granting it powers to expropriate land where necessary in the interests of the public. The following year, the rail network was nationalized.

1938: The property of foreign oil companies in Mexico was seized by the Government, and a national company, Petróleos Mexicanos (Pemex), was established.

1940: Manuel Avila Camacho was elected President; he revised many of the Government's anti-clerical policies.

22 May 1942: Following an attack on two Mexican tankers by German submarines, Mexico declared war on that country and its allies in the Second World War. Through a system of contract labour known as *bracero*, Mexican workers were employed to assist the US war effort in its south-western states.

January 1946: The PRN changed its name to the Partido Revolucionario Institutional (PRI); one of its members, Miguel Alemán Valdés, was subsequently elected President.

September 1947: Shortly after a visit by the President to the USA, the Government entered into the Inter-American Reciprocal Assistance agreement, which guaranteed a system of defence for western nations against outside aggression.

1952: Adolfo Ruíz Cortines was elected President; he instituted a number of policies aimed at eradicating corruption.

1953: Women were enfranchised.

August 1958: Adolfo López Mateos was elected President; he enacted a series of reforms concerning land, national health and labour relations.

1964: Gustavo Díaz Ordaz became President; his administration was beset by increasing unrest at economic and social problems.

20 October 1968: Police opened fire on student demonstrators at the Plaza of the Three Cultures in Tlateloco; the security forces had been instructed to suppress expressions of unrest during the country's hosting of the Olympic Games. A total of 325 demonstrators were killed, many injured and thousands were arrested.

1970: Luis Echeverría Alvarez was elected President; his administration promised to address the country's economic and social problems; he also sought to loosen the country's ties with the USA and instead cultivate relations with other Latin American countries.

July 1976: The PRI candidate, José López Portillo, won 95% of votes cast in a presidential election. The following year, he introduced legislation to increase the representation of minority parties in the legislature.

July 1982: Despite the participation of left-wing groups in presidential elections, the PRI candidate, Miguel del la Madrid Hurtado, was successful. The PRI also won the majority of seats in elections to the Cámara.

December 1982: The Government introduced a programme of economic reform, which in particular gave precedence to the repayment of the country's debts.

February 1985: Relations with the US Government deteriorated, following the murder of an agent of the US Drugs Enforcement Administration (DEA) by Mexican drugs traffickers.

July 1985: In congressional elections, the Government won 288 of the 300 directly elective seats in the Cámara.

1986: Dissidents within the PRI formed the Corriente Democrática (CD), led by Cuauhtémoc Cárdenas Solórzano.

March 1987: The Partido Mexicano Socialista (PMS) was founded.

July 1988: Congressional and presidential elections were held; a new coalition, the Frente Democrático Nacional (FDN), comprising the CD and four left-wing groups, participated. The FDN challenged the legitimacy of the PRI's victory; the opposition gained seats in the Senado and defeated the PRI in the Distrito Federal and three other states. A congressional committee was subsequently established to investigate allegations of electoral malpractice.

September 1988: In protest at the PRI's apparent obstruction of the investigations of the electoral commission, members of the FDN and other opposition groups withdrew from the Cámara. The PRI proclaimed Carlos Salinas de Gortari as the new President.

January 1989: The Government, trade unions and employers' organizations made a Pact for Economic Stability and Growth.

May 1989: The FDN was dissolved; members of the CD and the PMS founded the Partido de la Revolución Democrática (PRD), led by Cárdenas.

October 1989: Constitutional amendments to facilitate the adoption of a federal electoral code proposed by the PRI, were approved by both the Cámara and the Senado. The PRD criticized the inclusion of a 'governability' clause, which stipulated that the majority of seats in the Cámara should be awarded to the main party, should it receive at least 35% of votes cast in a general election.

April 1990: The Government condemned, as a violation of extradition agreements with the USA, the abduction of a Mexican physician in Mexico by agents working for the DEA , in connection with the murder of a DEA agent in 1985.

June 1990: Owing to increasing allegations of human rights abuses by the security forces, the President created the National Commission for Human Rights (CNDH); however, opposition groups criticized the exclusion of claims involving political campaigns and electoral processes from its mandate.

July 1990: A new electoral code was introduced; it provided for the issue of more detailed identification cards for voters, the compilation of a new electoral roll and the establishment of a new Federal Electoral Tribunal.

August 1990: The PRI won 290 of the 300 directly elective seats in the Cámara, in addition to 30 awarded by proportional representation. The party also gained 31 of 32 seats in the Senado.

January 1991: The Government signed a free trade agreement with Honduras, Guatemala, El Salvador, Nicaragua and Costa Rica.

January 1992: Following the introduction of legislation recognizing the Roman Catholic Church, the President revoked the constitutional restrictions on the church.

December 1992: The Government signed the North American Free Trade Agreement (NAFTA), which had earlier come into effect between the USA and Canada on 1 January 1989, providing for the reduction of tariffs between the countries.

July 1993: Amid continuing criticisms of the electoral process, the Government allowed parties access to the media and the 'governability' clause was removed.

January 1994: Indigenous insurgents, who had formed the Ejército Zapatista de Liberación Nacional (EZLN), took control of parts of Chiapas. They issued the Declaration of Lacandena Jungle, in which they listed their social and economic grievances, including their opposition to the country's participation in NAFTA, and declared war on President Salinas. A cease-fire was declared at the end of the month and the Government appointed Manuel Camacho Solís to lead a peace and reconciliation commission in Chiapas.

March 1994: The EZLN rejected the Government's offer to accelerate an anti-poverty programme in the region and reiterated their opposition to NAFTA, which they believed marginalized the interests of indigenous groups. Camacho Solís was subsequently replaced by Jorge Madrazo Cuéllar.

March 1994: The PRI presidential candidate, Luis Donaldo Colosia, was assassinated at a public rally. A suspect, Mario Aburto Martínez, was arrested at the scene.

July 1994: A report of the investigation into Colosia's murder claimed that Aburto Martínez had been motiveless and acted alone; he was sentenced to 42 years' imprisonment. Public condemnation of the report prompted President Salinas to order further investigation.

21 August 1994: Ernesto Zedillo Ponce de León of the PRI was elected President, gaining 48% of the votes cast; the Partido Acción Nacional (PAN) candidate, Diego Fernández de Cavallos, gained 25.9% and Cuauhtémoc Cárdenas of the PRD 16.6%. UN observers claimed that there had been instances of electoral malpractice.

September 1994: The PRI Secretary-General, José Francisco Ruiz Massieu (a former brother-in-law of outgoing President Carlos Salinas de Gortari), was murdered. His brother, Mario, subsequently resigned from his position of Deputy Attorney-General, alleging that the President had obstructed investigations into the murder.

October 1994: Owing to the presence of increasing numbers of government troops in the region, the EZLN announced its withdrawal from negotiations with the Government. The conflict in the Chiapas region escalated.

December 1994: Opposition groups in Chiapas disputed the election of the PRI's candidate, Eduardo Robledo Rincón, to the position of state governor and established their own parallel administration. A Comisión Nacional de Intermediación (CONAI) was subsequently established to mediate between the two sides.

February 1995: Following the discovery of illegal arsenals in Mexico City and Veracruz, the Government announced that it would launch a series of offensives against the EZLN. Later in month however, Robledo Rincón was recalled and the operations were suspended. In March, a Law for Dialogue, Conciliation and Honourable Peace in Chiapas was enacted by the Government.

February 1995: The Attorney-General, Antonio Lozano Gracia, concluded that the assassination of Colosia had involved two murderers and ordered the investigations to continue.

February 1995: Raúl Salinas de Gortari, the brother of the former President, was arrested on charges of Ruiz Massieu's murder.

April 1995: Fernando Rodríguez, who had been charged with hiring Ruiz Massieu's assassins, implicated several new conspirators; Mario Ruiz Massieu was subsequently detained in the USA on charges of corruption.

February 1996: Representatives of the EZLN and the Government met at San Andrés Larraínzar and concluded an accord, which guaranteed indigenous groups cultural, linguistic and local government rights.

July 1996: Eight alleged members of the guerrilla group, Ejército Popular Revolucionario (EPR), were arrested in Guerrero. Members of the group perpetrated attacks in México and Oaxaca the following month.

July 1996: The PRI, PAN, PRD and the Partido de Trabajo (PT) reached an agreement on electoral reforms, including the introduction of a directly-elected Governor of the Distrito Federal, an increase in financial aid to political parties, the introduction of proportional representation in elections to the Senate and the right of citizens resident abroad to vote.

September 1996: The EPR declared a cease-fire in Guerrero.

October 1996: Suspicions arose when a body found in the grounds of the home of Raúl Salinas was proved not to be that of Manuel Muñoz Rocha, a former federal congressman, considered to be Salinas' accomplice, as formerly believed.

December 1996: A multi-party congressional committee, the Comisión de Concordia y Pacificación (COCOPA), drafted legislation on the rights of indigenous groups, based on the San Andrés Larraínzar agreement; the document was rejected by the Government.

January 1997: The leader of the EPR, Benigno Guzmán Martínez, was arrested.

March 1997: Following the Government's extradition of Juan Garciá Abrego, the head of the Golfo drugs cartel, who had been accused of transporting illicit drugs to the USA, the US Government certified Mexico as a country co-operating in its campaign against drugs-trafficking.

May 1997: President Zedillo and US President William (Bill) Clinton signed agreements on border security in order to combat illegal trading in drugs and weapons.

6 July 1997: The PRI won 239 seats of the 500 seats available in elections to the Chamber of Deputies; the PRD and the PAN made significant gains. Cárdenas, as the PRD candidate, was elected to the position of Governor of the Distrito Federal.

September 1997: The Frente Zapatista de Liberación was inaugurated at a demonstration of Zapatistas and associated groups in Mexico City.

December 1997: Some 45 indigenous women and children were killed at a church in the village of Acteal, in Chiapas; the local mayor, Jacinto Arias, was among those arrested. Two years later, he was convicted of supplying weapons for the massacre and sentenced to 35 years' imprisonment.

June 1998: The armed forces killed 11 EPR members and arrested 21 others during a military offensive in Guerrero.

January 1999: Salinas was convicted of the murder of José Ruiz Massieu; his initial sentence of 50 years' imprisonment was later reduced to 28.

March 1999: A nation-wide unofficial plebiscite conducted by Zapatistas to gauge the level of public support for the improvement of indigenous rights, the San Andrés Larraínzar agreement and the demilitarization of Chiapas, indicated that 97% of the 2.5m. people questioned concurred with their demands.

September 1999: Mario Ruiz Massieu committed suicide in the USA, where he had been awaiting trial for charges of money-laundering; his suicide note blamed corruption in the PRI for his brother's and his own death.

October 1999: Jacobo Silva Nogales, a former EPR member and founder of the new Ejército Revolucionario Popular Insurgente group, was arrested.

November 1999: Soldiers and US Federal Bureau of Investigation agents excavated four suspected mass graves near Ciudad Juárez near the US border. They were thought to contain the bodies of at least 100 people who had belonged to the Juárez drugs cartel.

2 July 2000: In presidential elections, Vicente Fox Quesada, representing the Partido Verde Ecologista de México (PVEM)-PAN alliance, gained 43.5% of the votes cast, Francisco Labastida Ochoa of the PRI won 36.9% and Cárdenas of the Alianza por México secured 17.0%. This was the first time in Mexican history that a candidate of the ruling party had lost a presidential election. In concurrent congressional elections, the Alizana por el Cambio won a majority, securing 223 of the 500 seats available; the PRI however, maintained a majority in the Senate. Andrés Manuel López Obrador of the PRD won the governorship of the Distrito Federal.

January 2001: The new Government proposed that conflicting sides in Chiapas should open offices in the state's major town to facilitate negotiations. President Fox also pledged to reduce poverty and improve relations with the USA.

February 2001: President Fox introduced constitutional amendments, which endowed Congress with the right to impeach the President and approve cabinet appointments; guarantees of indigenous and human rights, in addition to measures to prevent electoral malpractice, were also introduced.

28 March 2001: Four EZLN members addressed Congress in preparation for the start of formal dialogue regarding a solution to the conflict in Chiapas.

29 April 2001: Legislation to safeguard indigenous rights was approved in the Congress, despite criticism from opposition groups that it was not effective enough. The bill was rejected by the Zapatistas, who claimed that it represented the rights of landowners rather than those of the indigenous peoples.

25 May 2001: The former state Governor of Quintana Roo, Mario Villanueva, was arrested following his disappearance two years previously in connection with drugs charges; it was alleged that he had assisted drug cartels to import large quantities of cocaine during his period in office.

7 September 2001: During a visit to the USA, President Fox suggested to the US President, George W. Bush, that the Inter-American Treaty of Reciprocal Assistance should be revoked and replaced by a new agreement aimed at combating poverty and organized crime in the region. He also urged the US Government to naturalize millions of Mexicans who had entered the USA illegally.

Nicaragua

c. **AD 800:** The Chorotega people migrated south from Mesoamerica.

c. **1200:** Aztec peoples migrated south and established a trade route that linked Mexico and Guatemala with South America.

1502: An Italian navigator in the service of the Spanish Crown, Cristoforo Colombo (Christopher Columbus), landed on the Caribbean shores of Central America.

1522: Nicaragua was named after the local Niquirano chief by the Spanish explorer Gil González de Avila.

1523–24: The conquest of Nicaragua was completed by the Spanish navigator Francisco Hernández de Córdoba, who founded the towns of Granada and León.

15 September 1821: The Central American provinces, including Nicaragua, declared their independence from Spain, but Nicaragua briefly came under Mexican rule.

1823: The United Provinces (Federation) of Central America, comprising Costa Rica, El Salvador, Guatemala, Honduras and Nicaragua, was established.

30 April 1838: Nicaragua seceded from the United Provinces of Central America and gained full independence.

1860: The British ceded control of the Caribbean coast (which they had controlled since the 17th century) to Nicaragua.

1893: Gen. José Santos Zelaya seized power and established a military dictatorship.

1909: Gen. Zelaya was deposed with the assistance of US armed forces.

1912–25: US military bases were established at the request of the Nicaraguan Government.

1927-33: Guerrilla groups led by Augusto César Sandino campaigned against the US military presence. However, following the creation of a National Guard (commanded by Gen. Anastasio Somoza García), US troops departed from Nicaragua.

21 February 1934: Sandino was assassinated by the National Guard, but his supporters ('Sandinistas') continued to oppose the regime.

1 January 1937: Gen. Somoza officially became President, having seized power in a *coup d'état* in 1935.

29 September 1956: President Somoza died from wounds inflicted during an assassination attempt and was replaced by his son, Luis Somoza Debayle.

23 July 1961: The Frente Sandinista de Liberación Nacional (FSLN) was founded as a left-wing opposition front which sought the overthrow of the Somoza family by revolution.

1 June 1967: Following the death of President Luis Somoza, his brother, Gen. Anastasio Somoza Debayle, became President.

April 1972: The Congreso Nacional was dissolved. Nicaragua was ruled by a triumvirate from 1 May until Gen. Somoza was re-elected as President in September 1974.

10 January 1978: The assassination of Pedro Joaquin Chamorro, leader of the opposition coalition, provoked violent protests against the Government.

16 June 1979: The FSLN announced the creation of a provisional Junta of National Reconstruction.

17 July 1979: With the FSLN preparing to seize control of the capital, Managua, President Somoza was forced into exile in Paraguay (where he was assassinated in September 1980).

20 July 1979: The Junta and its Provisional Governing Council took power as the Government of National Reconstruction. The existing Constitution was abrogated, the Congreso Nacional dissolved and an appointed Council of State was to act as an interim legislature. The National Guard was disbanded and replaced by the Ejercito Popular Sandinista (EPS).

4 March 1981: Cmdr Daniel Ortega Saavedra was appointed Co-ordinator of the Junta and of its new consultative body, the Council of Government.

March 1982: The Sandinista Government declared a state of emergency, after US-sponsored counter-revolutionary forces ('Contras'), mostly composed of members of the former National Guard, began to operate from camps in Honduras.

June 1984: Talks commenced between the Nicaraguan and US Governments in order to foster peace initiatives proposed by the 'Contadora group' (Colombia, Mexico, Panama and Venezuela).

4 November 1984: The FSLN candidate, Daniel Ortega, was elected President. In elections to the new National Constituent Assembly, the FSLN won a majority.

July 1985: Miskito Indians began to return to their ancestral homes in northern Nicaragua, following talks with the Government concerning regional autonomy.

9 January 1987: A new Constitution came into effect.

17 August 1987: A peace plan for Nicaragua was signed by the five Central American states. A National Commission for Reconciliation was formed and, in January 1988, the Government ended the state of emergency and agreed to negotiate with the Contras.

23 March 1988: A 60-day cease-fire was agreed, although this was later extended unilaterally by the Government until November 1989. The Government agreed to release political prisoners and to permit the Contras to participate in future elections.

June 1989: The Unión Nacional Opositora (UNO) was formed by 14 opposition parties.

7 August 1989: The five Central American Presidents signed an accord providing for the voluntary demobilization, repatriation or relocation of the Contras within 90 days. However, in November President Ortega declared an end to the cease-fire, claiming that the rebels were still operating from bases in Honduras.

25 February 1990: The UNO defeated the FSLN in legislative elections and its candidate, Violeta Barrios de Chamorro (the widow of Pedro Joaquín Chamorro), defeated Ortega to be elected President. The UNO failed to secure a sufficient number of seats in the Asamblea Nacional to amend the Constitution and radically reform the EPS.

March 1990: The outgoing FSLN Government introduced the *piñata* laws, which guaranteed property rights for the thousands of Nicaraguans who had benefited from the land expropriation conducted by the Sandinistas.

19 April 1990: The EPS and the Contras agreed the terms of a cease-fire.

25 April 1990: The Chamorro administration assumed power and began to reverse many Sandinista policies. However, widespread public unrest forced the UNO to concede to many of the FSLN's demands.

27 June 1990: The demobilization of the Contras was officially concluded, signalling the end of Nicaragua's 11-year civil war.

18 June 1991: The FSLN withdrew indefinitely its 39 deputies from the Asamblea Nacional, in protest against proposals to revoke the *piñata* laws concerning the redistribution of property.

21 July 1991: The first congress of the FSLN appointed Daniel Ortega to the newly-created post of General Secretary.

23 August 1991: The legislature approved the abrogation of the *piñata* laws, although on 11 September President Chamorro vetoed parts of the bill that she deemed to be unconstitutional.

29 August 1991: A National Security Commission (including a 150-strong Brigada Especial de Desarme—BED) was established in order to disarm civilians. However, hostilities continued between rearmed Sandinistas ('Re-compas') and rearmed Contra rebels ('Re-contras'), who claimed that the Contras had not received the land and aid pledged to them under the resettlement plan.

April 1992: Although the phased disarmament of the Re-contras and Re-compas had begun in January, groups of the former combatants joined forces to form the 'Revueltos', demanding land and credit promised to them prior to demobilization.

June 1992: The Government secured a court injunction which prevented a legislative vote of censure being taken against Gen. Humberto Ortega, the Chief of the EPS, who was accused of maintaining a clandestine pro-Sandinista army.

9 September 1992: In response to US and UNO criticism on the issue of the return of property expropriated under the Sandinista Government, President Chamorro signed three decrees establishing a property ombudsman's office and an agreement to expedite the processing of property claims.

30 December 1992: After the President of the Asamblea Nacional had, in September, convened the legislature in the absence of the FSLN deputies and a group of dissident UNO deputies called the Grupo de Centro (GC), and had recruited substitute deputies to elect new legislative authorities, President Chamorro ordered the army to occupy the assembly building and appointed a provisional administration to manage legislative affairs pending the election of new authorities.

30 January 1993: New legislative authorities were appointed and a new Cabinet formed. Without a parliamentary majority or representation in the Cabinet, the UNO declared itself as an opposition party, the Alianza Política Opositora (APO), expelling four member parties for their involvement with the GC.

May 1993: The Government held separate talks with the APO, the FSLN and trade unionists in an attempt to establish the agenda for a national dialogue. However, the APO demanded that its parliamentary majority be restored and Gen. Humberto Ortega be replaced as Chief of the EPS.

21–22 July 1993: The Government deployed more than 2,000 troops to quell an uprising by Re-compas in the northern town of Estelí; some 45 people (including several civilians) died in the ensuing violence.

17 August 1993: The Asamblea Nacional approved an amnesty for all political crimes.

19 August 1993: A Re-contra group, the Frente Norte 3-80 (FN 3-80), seized 38 hostages (BED members who had been sent to discuss the amnesty proposals with the rebels) in Quilalí and demanded government changes. In response, a group of Re-compas, the Comando 40 por la Dignidad de la Soberanía, seized the offices of the APO in Managua and took 33 hostages, including the APO's political council. By 26 August the hostages had been released and a temporary truce was declared between the Government and the FN 3-80, pending the group's disarmament.

15 October 1993: More than 2,500 government troops were deployed against the FN 3-80, which had failed to disarm as scheduled.

25 November 1993: Following unprecedented talks held between the two blocs, the APO and the FSLN signed an agreement providing for the implementation of partial constitutional reforms.

December 1993: Divisions within the APO resulted in the Unión Demócrata Cristiana (UDC), the Movimiento Democrático Nicaragüense (MDN) and the Alianza Popular Conservador (APC) breaking away from the alliance.

10 January 1994: Following the election of new legislative authorities, the APO abandoned its boycott of the Asamblea Nacional.

February 1994: Daniel Ortega founded the Izquierda Democrática Sandinista, an internal faction of the FSLN which was intended to maintain the party's role as a revolutionary force.

24 February 1994: Following a cease-fire agreed in early February, a peace settlement was signed by the Government and the FN 3-80, whereby the group would demobilize in exchange for being granted an amnesty and the right to be incorporated into the national police force.

April 1994: The APO was reduced to only four parties, following the departure of the Partido Liberal Constitucionalista (PLC, the Partido Social Demócrata having withdrawn from the alliance in February).

22 May 1994: At an extraordinary congress held by the FSLN, the revolutionary faction secured control of the party's national directorate and Daniel Ortega was re-elected as General Secretary (as he was again in May 1998).

23 August 1994: The Asamblea Nacional passed legislation that sought to depoliticize the armed forces and render them fully accountable to the civilian authorities.

24 November 1994: The Asamblea Nacional approved amendments to some 67 of the Constitution's 202 articles. The emphasis of the reforms was to adjust the balance of authority in favour of the legislature. The amendments were deemed illegal by the FSLN leadership. The EPS was renamed the Ejército de Nicaragua.

15 January 1995: The renewalist faction of the FSLN, led by Sergio Ramírez Mercado, founded a separate political party, the Movimiento Renovador Sandinista (MRS).

21 February 1995: Gen. Humberto Ortega was officially succeeded as Commander-in-Chief of the armed forces by Gen. Joaquín Cuadra Lacayo.

24 February 1995: The amendments to the Constitution were enacted by the legislature, although President Chamorro had refused to promulgate them. Chamorro condemned the decision as unconstitutional and contested the legitimacy of the reforms.

June 1995: Amid frequent disputes concerning fishing rights in the Gulf of Fonseca (in the Caribbean Sea), Honduras and Nicaragua signed an agreement providing for the visible demarcation of their territorial waters.

15–22 June 1995: The dispute between the Government and the Asamblea Nacional was ended by the signing of two political agreements, under the terms of which many of the amendments to the Constitution intended to reduce presidential authority were to be moderated.

4 July 1995: Legislation defining the interpretation and implementation of the constitutional amendments was approved by the Asamblea Nacional and subsequently endorsed by President Chamorro.

November 1995: Following protests by trade unionists affiliated to the FSLN, legislation was introduced legalizing the land titles of more than 200,000 families who had been granted land under Sandinista agrarian reform.

20 October 1996: In the presidential election, Arnoldo Alemán Lacayo of the PLC, who was representing the new Alianza Liberal, was initially said to have secured the share of the vote necessary for outright victory. However, amid claims of electoral fraud by the FSLN and other parties, many results were revised. Final results were announced on 22 November, when Alemán was declared President-elect, with some 51% of the votes, compared with Daniel Ortega's 38%. In the legislative elections, also held on 20 October, the Alianza Liberal failed to gain a majority in the Asamblea Nacional, winning 42 of the 90 seats. The FSLN obtained 35 seats, while nine smaller parties received small numbers of seats.

April 1997: The FSLN organized widespread protests in opposition to the Government's policy concerning the return of property expropriated under the Sandinista regime.

June 1997: A 'national dialogue', involving representatives of more than 50 political parties and civic organizations, was convened by the Government, with a view to achieving a consensus on the resolution of issues including poverty, unemployment and property rights. However, the FSLN declined to participate in the dialogue.

26 November 1997: A new law, the *Ley de la Propiedad Reformada, Urbana y Agraria*, was approved, whereby occupants of small expropriated properties would be granted legal ownership of the land, while those of larger properties would be required to return them or to compensate fully the owners over a 15-year period.

April 1999: Security forces were deployed in Managua to disperse violent demonstrations by students and transport workers, who were supported by the FSLN.

June 1999: The Government and the FSLN reached an accord to initiate negotiations on constitutional reform. This co-operation prompted the formation within the FSLN of a dissident anti-Ortega faction, which (in alliance with former Contras and various right-wing parties) protested against the pact.

10 November 1999: The Comptroller-General, Agustín Jarquín (who had launched numerous investigations into corrupt practices by the Alemán administration, citing the President in particular), was arrested on charges of committing fraud against the State. However, in late December Jarquín was acquitted of the charges and he formally renewed his corruption allegations against President Alemán.

30 November 1999: Nicaragua was angered when Honduras ratified the Caribbean Sea Maritime Limits Treaty (which it had signed with Colombia in 1986), which formally recognized a frontier with Colombia that effectively granted Colombia territorial rights to areas of the Caribbean Sea historically claimed by Nicaragua. Troop movements were subsequently reported on both sides of the Nicaragua–Honduras border.

8 December 1999: Following further confrontations (including exchanges of fire) between Honduran and Nicaraguan naval patrols, Nicaragua initiated proceedings at the International Court of Justice (ICJ) to determine the maritime delimitation in the Gulf of Fonseca.

January 2000: Following mediation by the Organization of American States (OAS), Nicaragua ended its trade sanctions against Honduras, which had been imposed in the previous month.

20 January 2000: A series of constitutional reforms came into force. The changes, which included electoral reforms and the restructuring of the judiciary, increased the likelihood of an FSLN election victory; however, in return, Alemán was rendered virtually immune from prosecution after his term of office had ended.

7 March 2000: Nicaragua and Honduras signed an agreement committing them to observe a maritime exclusion zone in the Caribbean Sea and to reduce troop numbers on their common border. Both countries agreed to submit all issues pertaining to maritime space in the Caribbean to the ruling of the ICJ, and to mount co-ordinated naval patrols in the Gulf of Fonseca.

late June 2000: The former Commander-in-Chief of the armed forces, Gen. Joaquín Cuadra Lacayo, founded the Movimiento de Unidad Nacional (MUN). In late July the former Minister of National Defence, José Antonio Alvarado, resigned from the PLC to form a new party, the Partido Liberal Democrático (PLD). However, in January 2001 both parties were refused official recognition by The Consejo Supremo Electoral.

5 November 2000: The PLC won a convincing victory in municipal elections held throughout Nicaragua; however, the FSLN secured control of several municipalities, including Managua.

27 March 2001: Amid claims that Honduras had violated the March 2000 accords by carrying out military exercises in the Gulf of Fonseca, Nicaragua submitted documentation to the ICJ contesting the Caribbean Sea Maritime Limits Treaty. Nicaragua also refused to participate in the joint Honduran–Nicaraguan naval patrol.

May 2001: Nicaragua and the USA re-established full military relations (suspended since 1980).

4 November 2001: Provisional results from the presidential elections showed that Enrique Bolaños, of the ruling PLC, had comprehensively defeated the former Sandinista President, Daniel Ortega.

Panama

AD 1502: The Spanish explorer, Rodrigo de Bastidas, visited what is now known as Panama, which was inhabited by Cuna, Choco, Guaymi and other indigenous Indian groups.

1513: The Spaniard, Vasco Núñez de Balboa, was made Governor of Panama, but was later executed.

1519: Panama became part of the Spanish viceroyalty of New Andalucia (later New Granada).

1821: Panama achieved independence from Spain and joined the confederacy of Gran Colombia, which included Colombia, Venezuela, Ecuador and Peru.

1830: Panama became part of Colombia following the collapse of Gran Colombia.

1847: The Panama Railroad Company was formed by a group of New York financiers; the company secured an exclusive concession from Colombia allowing construction of a track crossing from the Atlantic to the Pacific coast.

1850–55: The Panama Railroad Company undertook construction of a railway across Panama.

1879: Work began on building a canal to connect the Atlantic and Pacific oceans by the French Panama Canal Co.

1888: The attempt by the French Panama Canal Co. to build a canal failed as a result of financial difficulties and the death of thousands of workers from yellow fever and malaria.

November 1903: Panama achieved independence from Colombia. Earlier that year the USA had purchased the rights to build the Panama Canal from Colombia and was given control of the Canal Zone (a 10-mile-wide strip of Panamanian territory) for a 100-year period. However, objections to the treaty had been raised by the Colombian Congress, causing revolts in Panama.

February 1904: The US Senate ratified a treaty with the Panamanian Government, granting Panama sovereignty and the USA 'sovereign rights'. The USA effectively retained the right to military intervention to protect the Canal, and had its own military bases, police force and laws within the Canal Zone.

1939: Panama ceased to be a US protectorate by mutual agreement.

1941: The President of Panama, Arnulfo Arias Madrid, was overthrown in a bloodless *coup d'état* and replaced by Adolfo de la Guardia.

1945: De la Guardia refused to resign at the expiry of his term of office and was replaced by Enrique Jiménez.

1948: Domingo Díaz Arosemena defeated Arias Madrid at presidential elections, amid allegations of fraud on both sides.

1949: President Díaz died, and was replaced by Arias Madrid following a bloodless *coup d'état* led by Col José Antonio Remón, the former chief of police.

1950: President Arias Madrid was impeached and banned from public life, following a failed attempt to suspend the right of habeas corpus.

1952: Remón was elected President and set about enacting a programme of moderate reforms.

January 1955: Remón was assassinated.

23 January 1955: The Treaty of Mutual Understanding and Co-operation was signed with the USA. Under its provisions, US commercial activities not essential to the operation of the canal were to be reduced and a standard basic wage scale was implemented. Panama made concessions to the USA by allowing military bases to be constructed outside the Canal Zone, and leasing parts of its territory, to be used for US military manoeuvres, for no fee for a 15-year period.

May 1958: Students demonstrating against the USA and its policy of paying US citizens in Panama 25% more than local workers, clashed with the National Guard. Nine people were killed.

1968–81: Arias Madrid won a presidential election, however, Col (later Brig.-Gen.) Omar Torrijos Herrera, the chief of the National Guard, overthrew the elected President after only 11 days in office and imposed a dictatorship; freedoms of the press, of speech and assembly were suspended for one year and a programme of economic modernization was launched.

February 1969–October 1978: Party political activity was banned.

1972: Torrijos took the executive title of Chief of Government and legislative power was vested in the 505-member Asamblea Nacional de Representantes de Corregimientos.

September 1977: The USA and Panama signed two draft treaties agreeing to transfer the Canal to Panama as from 31 December 1999; a phased withdrawal of troops, with Panama eventually taking control of all US military bases in the Canal Zone (which would be renamed the Canal Area), was to commence prior to that date. The Panama Canal Co would be replaced by the Panama Canal Commission, and Panama and the USA were to be jointly responsible for guaranteeing the neutrality of the Canal.

October 1977: The treaties on the transfer of the Canal were approved in a national referendum.

1978: Torrijos announced plans to return Panama to elected government; he resigned as Chief of Government in October (retaining the post of National Guard Commander), when a newly elected Asamblea Nacional endorsed his nominee, Dr Arístides Royo Sánchez, as President for a six-year term. Torrijos maintained his hold on power however, when his party, the Partido Revolucionario Democrático (PRD) won 10 seats in the 19-seat Consejo Nacional de Legislación in August. The elections were boycotted by Arias Madrid's Partido Panameñista Auténtico (PPA).

October 1979: Following a two-year delay, the US Congress ratified the treaties on the transfer of the Canal.

1981: Torrijos died in an air crash, which was claimed to be accidental, although allegations were made about US Central Intelligence Agency (CIA) involvement.

March 1982: Gen. Rubén Darío Paredes, a keen advocate of pro-US foreign policy, was appointed as Commander of the National Guard.

July 1982: Gen. Paredes forced Royo from office, and the First Vice-President, Ricardo de la Espriella, became President.

August 1983: Paredes withdrew from the National Guard in order to contest the 1984 presidential election, and was replaced by former intelligence chief and one-time CIA informant, Brig. (later Gen.) Manuel Antonio Noriega Morena; he increased the size of the force, which he renamed the Panama Defence Forces, and greatly increased its power over Panama's political and economic life.

February 1984: Dr Jorge Illueca took over as President.

May 1984: Nicolás Ardito Barletta of the PRD was elected President by a narrow margin.

September 1985: Barletta resigned, amid allegations that the Defence Forces had assisted his election; he was replaced by Eric Arturo del Valle, who pledged a return to 'Torrijista principles', but faced strong opposition to his economic policies.

1987: Gen. Noriega, the effective ruler since 1983, resisted calls for his removal, despite the suspension of US military aid and the imposition of economic sanctions.

February 1988: Del Valle was replaced by Manuel Solis Palma following an attempt by him to depose Noriega. The latter, charged with drug smuggling by the USA, declared a state of emergency after the coup against him failed.

May 1989: Noriega declared the results of legislative elections to be invalid after they were won by the opposition.

December 1989: Noriega was declared Head of Government; later that month he declared a 'state of war' with the US, and on 20 December US forces intervened to overthrow him; Noriega was deposed.

21 December 1989: Guillermo Endara, who had won the disputed May elections, was installed as President. US economic sanctions were lifted.

February 1990: A new Asamblea Legislativa was formed, based on the results of the May 1989 elections. The Alianza Democrática de Oposicion Civilista (ADOC) won 51 seats, the Coalición de Liberación Nacional (COLINA) took six and new elections were held for the remainder.

October 1991: An attempted *coup d'état* against the Government was unsuccessful.

26 December 1991: Constitutional reforms were approved by the Asamblea, including the abolition of a standing army; a privatization programme was introduced.

10 July 1992: A US court found Noriega guilty of drug offences and money laundering and sentenced him to 40 years' imprisonment, to be served in a US prison.

15 November 1992: A referendum proposing the constitutional reforms approved by the Asamblea in December 1991 was defeated.

8 May 1994: Ernesto Pérez Balladares of the PRD was elected President; the Constitution was amended by the Asamblea and the army was formally abolished.

4 October 1994: Noriega was sentenced *in absentia* by a Panamanian court to 20 years' imprisonment for the murder of a senior military officer in 1989.

30 August 1998: Voters rejected a proposed constitutional amendment allowing the President to seek election to a second term.

March 1999: Noriega's sentence was reduced to 30 years, following an appeal by his lawyers claiming that he had given years of service to the USA as an 'asset' of the CIA.

2 May 1999: Nueva Nación (NN), led by Martín Torrijos, the son of Gen. Torrijos, won a majority at legislative elections, taking 41 of the 71 seats; the Unión por Panama (UPP) won 24.

1 September 1999: Mireya Moscoso of the UPP became Panama's first female President, defeating Martín Torrijos of the NN. On the same day a Government of National Unity was formed, comprising members of the Partido Solidaridad, the Partido Liberal Nacional (PLN), the Partido Demócrata Cristiano (PDC) and the Partido Renovación Civilista (PRC). Moscoso reached an agreement with six minority parties in order to achieve a working parliamentary majority of one seat.

November 1999: The last US military base in the Canal Area was closed.

31 December 1999: Panama took full control of the Panama Canal, ending nearly a century of US jurisdiction over one of the world's most strategic waterways. The canal became a fully commercial operation (under US control it had been run on a non-profit basis).

March 2000: A five-year modernization project for the Canal was announced, including the development of technology to raise capacity, the construction of a second bridge and the widening of the narrowest section of the Canal.

24 August 2000: The PRD and the PDC formed an alliance, removing the governing coalition's small majority.

27 December 2000: Moscoso announced the creation of a panel to investigate crimes committed while military governments were in power between 1968 and 1989.

Paraguay

AD 1524: The country, inhabited by the Guaraní peoples, was visited by the Portuguese explorer Alejo García, in the service of Spain.

1530s: The Italian navigator Sebastiano Caboto, also in the service of Spain, partly explored the rivers of the region, chiefly the Paraná.

15 August 1537: Spanish adventurers seeking gold established a fort on the Paraguay River, calling it Nuestra Señora de la Asunción, after the Christian feast celebrated that day. The city, commonly known as Asunción, eventually became the centre of a Spanish colonial province, although dependent on the city of Buenos Aires.

1609: Jesuit Roman Catholic missionaries began to establish *reducciones*, settlements of indigenous converts to Christianity, whom the missionaries educated.

1620: The territory became a separate dependency of the Viceroyalty of Peru.

1767: The Jesuits were expelled.

1776: Spain created the Viceroyalty of the Río de la Plata, which comprised present-day Paraguay, as well as much of Argentina, Bolivia and Uruguay.

July 1810: Two months after the people of Buenos Aires had declared self-government, a group in Asunción took similar action; repudiating both the colonial administration and the leadership in Buenos Aires.

January 1811: A Paraguayan army defeated and drove back the forces of Gen. Manuel Belgrano, sent to compel Paraguay's adherence to Buenos Aires' leadership.

14 May 1811: A congress in Asunción declared Paraguay's independence. A five-man junta was elected to rule the new nation with Dr José Gaspar Rodríguez de Francia declared Consul.

12 February 1814: Fulgencio Yegros became Consul.

12 June 1814: De Francia returned as Consul. Congress voted to give full powers of rule by decree for three years, shortly thereafter extending this to a life term. De Francia pursued an isolationist policy, closing the country's borders and declaring Guaraní to be the sole official language.

September 1840: Dr Francia died; a succession of leaders were unable to establish adequate authority.

March 1841: Carlos Antonio López, a nephew of de Francia, assumed power and was granted full executive authority. He reversed the isolationist policy, encouraged commerce, instituted many administrative reforms, and began the construction of a railway.

1845: President López became involved in altercations with the Argentine military leader, Juan Manuel de Rosas, and joined the coalition against Buenos Aires.

1862: President López died; he was succeeded as President-for-life by his son, Francisco Solano López.

1864: Paraguay went to war against the combined forces of Argentina, Uruguay, and Brazil; the conflict became known as the War of the Triple Alliance.

1870: Paraguay was defeated when Brazilian soldiers executed President López after capturing him in battle. Argentina and Brazil annexed some 140,000 sq km of Paraguayan territory and levied a large indemnity payment. The war, known in Paraguay as the 'National Epic', is estimated to have resulted in the deaths of some 90% of the country's adult male population.

1870: Following disputes over control of the defeated country, the Brazilian-supported Asociación Nacional Republicana—Partido Colorado took power; internal disputes within the Colorados prevented any President from serving a full term of office before being deposed, although the party nominally remained in power until 1904. Between 1870 and 1979 the country had 45 Presidents, all of whom achieved power by force or amid allegations of electoral irregularities.

1883: Legislation provided for the sale of land previously accessible to all Paraguayans; the majority of such common land was assumed into large estates and those who had previously worked the land were forced either to leave or to become labourers.

1904: A popular uprising brought the Partido Liberal to power, led by Juan Bautista Gaona. The political instability remained however, with few Presidents able to serve their full term of office.

1907: Bolivia and Paraguay enlarged their frontier garrisons in response to the continued dispute regarding the sovereignty of the Chaco region, which was rumoured to be the site of substantial petroleum deposits, and into which both countries had expanded their control.

1927: The Argentine President, Máximo Torcuato de Alvear, made an unsuccessful attempt to secure an agreement between Bolivia and Paraguay over the Chaco territory.

December 1928: Paraguayan patrols attacked a Bolivian fort at Vanguardia, the first of a series of border incidents in the Chaco.

1932: Military tensions in the Chaco escalated into conflict, which became known as the Chaco War. Paraguay gradually gained the ascendancy.

1935: A cease-fire was signed by Paraguay and Bolivia. Rival groups of war-veterans effected two *coups d'état* in the subsequent two years.

1938: In the final settlement of the Chaco War, made by an arbitration commission under Argentine influence, Paraguay was granted about three-quarters of the disputed area.

1940: A new Constitution was adopted, vesting in a centralized government the power to regulate economic activity.

September 1940: On the death of the President, Marshal José Félix Estigarribia (the Liberal leader in the Chaco War) in an aeroplane crash, his former Minister of War, Gen. Higinio Morínigo, took power. Morínigo banned all political parties and severely repressed the activities of the trade-union movement. The Febreristas, as the followers of Col Rafael Franco (deposed by Marshal Estigarribia in 1937), were known, were suppressed and prominent figures within the movement were exiled.

1946: The Allied victory in the Second World War gave impetus to a military uprising against President Morínigo, whose sympathies were with the opposing Axis powers. Morínigo was forced to accept a Colorado-Febrerista coalition Government under his nominal control.

March 1947: Civil war escalated when Morínigo excluded the Febreristas from the Government and openly sided with the Colorados. The intervention of the Argentine Government of Gen. Juan Domingo Perón Sosa ensured a Colorado victory over the Liberals, Febreristas and Communists, who were led by Col Franco.

June 1948: A *coup d'état* led to Morínigo's removal from power. The sole candidate at the resultant presidential election was the Colorado, Juan Natalicio González; the Colorados subsequently became the only party represented in the legislature—Paraguay thus became a one-party state, although factionalism within the Colorados reduced the movement's effectiveness.

September 1949: Dr Federico Chávez assumed the presidency; his regime, authoritarian in nature, resembled that of Morínigo. Chávez imposed a state of siege three weeks after he took office, aiming his emergency powers at supporters of González, who eventually fled into exile.

October 1951: Gen. Alfredo Stroessner Mattiauda, a veteran of the Chaco War, was appointed Commander-in-Chief of the Armed Forces.

4 May 1954: Amid a deteriorating economic situation, Gen. Stroessner ordered his troops to depose the Government after President Chávez had attempted to dismiss one of his subordinates. Forces loyal to President Chávez resisted the attack, but were eventually defeated, and Gen. Stroessner assumed power.

July 1954: After being nominated as the Colorado candidate for the presidency, Stroessner was the sole candidate in the election to remainder of Chávez's term in the presidency.

15 August 1954: Gen. Stroessner was inaugurated as President, representing, to many members of the divided Colorados an 'interim' choice. President Stroessner declared a state of siege, which he renewed at intervals prescribed by the Constitution, and assumed extensive executive powers. Restrictions on political activity were enforced and Febrerista and Liberal groups suppressed.

November 1955: Epifanio Méndez Fleitas, the former head of the central bank who had assisted President Stroessner in the planning of his *coup d'état*, prepared to seize power himself in late December. However, President Stroessner pre-empted the insurrection, purging the military of Méndez Fleitas's supporters; Méndez Fleitas himself was forced into exile.

1956: An agreement was signed with Brazil to improve the transport link between the two countries by building roads and a bridge over the Paraná. This agreement marked the end of Paraguayan dependence on Argentine goodwill for the unhindered entry and exit of Paraguayan international trade.

1958: A national plebiscite elected Stroessner to a second presidential term, although dissatisfaction with the regime provoked the creation of a number of guerrilla groups. Sponsored by exiled Liberals and Febreristas, small bands of armed men, aided by Venezuela, began to enter the country from bases in Argentina.

1959: The Cuban revolutionary Government led by Dr Fidel Castro Ruz provided assistance to the guerrillas, whose attacks increased in frequency and severity.

April 1959: President Stroessner announced a programme of reforms of the Army and the Colorado Party. He lifted the state of siege, allowed opposition exiles to return, ended press censorship, freed political prisoners, and committed to a redraft of the 1940 Constitution.

May 1959: Some 100 people were injured in rioting in Asunción. The disturbance inspired the legislature to demand the resignation of the Minister of the Interior, Edgar Ynsfrán. President Stroessner responded swiftly by re-imposing the state of siege and dissolving the legislature. Some 400 Colorado politicians who opposed President Stroessner were imprisoned or fled into exile where they formed the Movimiento Popular Colorado (MOPOCO), under the leadership of Méndez Fleitas.

1960–61: Sporadic guerrilla incursions were launched by dissident Liberals belonging to the Moviemiento 14 de Mayo and by Communist groups.

1960s: A non-violent peasant movement, the *ligas agrarias*, emerged to campaign against the system of land tenure and in favour of agrarian reform; it was encouraged by the reform movement within the Roman Catholic Church. The Government accused the Church of fomenting Communism.

1963: In return for the participation of the Renovation Movement, a faction of the Liberal Party, in the elections of that year, Stroessner allotted the new party 20 of the 60 seats in Congress.

May 1966: A police-corruption scandal concentrated public opposition to Ynsfrán; President Stroessner dismissed him in November. A subsequent purge of the Colorado Party eliminated all internal opponents of the President.

August 1967: A new Constitution created a bicameral legislature and formally allowed Stroessner to serve for two further five-year presidential terms.

1968: Members of the Partido Liberal Radical (PLR), the remnants of the Liberal party, returned to Paraguay from exile and began participating in the electoral process (the Febreristas, by now a weak political force, had been legalized in 1964). The new Partido Demócrata Cristiano also renounced violence as a means of gaining power. The exhaustion of most opposition forces enabled President Stroessner to suppress the Partido Comunista Paraguayo and to isolate the exiled Colorado *epifanistas* (followers of Méndez Fleitas) and the MOPOCO.

1969: Relations between the Roman Catholic Church and the state deteriorated sharply when Church property was invaded by the police. The Archbishop of Asunción, Aníbal Mena Porta, refused to participate in the Council of State.

1973: Stroessner was re-elected to the presidency.

1976: The Government accused the *ligas agrarias* of being linked to an embryonic guerrilla movement among left-wing students, the Organización Primero de Marzo. The accusation was used as justification for repressing peasant organizations; some 50 peasant leaders were killed, several hundred fled into exile and more than 5,000 were arrested.

1977: Domingo Laíno, a PLR congressman, left the party to form the Partido Liberal Radical Auténtico (PLRA). Laíno accused the Government of corruption, involvement in drugs-trafficking, human-rights violations, and inadequate financial compensation from Brazil with regard to the Itaipú dam, constructed on the Paraná for power-generation purposes and financed by Brazil.

1978: Stroessner was re-elected to a fifth term as President. The Inter-American Commission on Human Rights persuaded an annual meeting of foreign ministers at the Organization of American States (OAS) to adopt a resolution urging Paraguay to improve its human-rights situation.

1979: The Febreristas, MOPOCO, the PDC and the PLRA formed an alliance, the Acuerdo Nacional (AN), which served to co-ordinate the opposition's political strategy—the impetus for the alliance had been President Stroessner's decision to seek a further term of office, thus requiring further constitutional amendments. Stroessner was re-elected later in the year.

1980: The Ninth OAS General Assembly in La Paz (Bolivia), condemned human-rights violations in Paraguay.

August 1980: Antonio Maidana, the leader of the PCP, was abducted in Buenos Aries and transferred to Paraguay, where he subsequently disappeared.

June 1981: Luis Alonso Resck, the leader of the PDC, was deported.

December 1982: Laíno was forced into exile.

February 1983: Stroessner was re-elected to the presidency. Opposition parties boycotted the poll, and the accompanying legislative elections.

December 1983: In an attempt to foment divisions within the AN, President Stroessner allowed MOPOCO leaders to return to Paraguay after nearly 25 years in exile, causing speculation that the Partido Colorado would be reunified. However, the MOPOCO leaders remained loyal to the Acuerdo Nacional.

March 1984: President Stroessner closed *ABC Color*, the country's most widely circulated newspaper, and arrested its owner. The newspaper had criticized the Government's handling of the Itaipú treaty with Brazil and denounced corruption in the state petroleum refinery, Repsa.

May 1984: President Stroessner celebrated 30 years in office.

March 1985: An estimated 4,000 demonstrators at a PRLA rally in Caraguatay clashed with police who attempted to suppress the protest.

May 1985: Police blockaded the capital's main public hospital for three days in response to demonstrations by staff. Subsequently, an independent trade-union body, the Movimiento Intersindical de Trabajadores (MIT) and an independent peasant body, the Movimiento Campesino Paraguayo (MCP) became the focus of popular opposition to the regime.

September 1985: A radical faction of the Colorado Party, the *eticos*, formed around Carlos Romero Pereira. The new movement denounced corruption and openly demanded a dialogue between the Government and the opposition parties of the AN. In response, the traditionalist Colorado faction, expressed support for Juan Ramon Chaves, who had been the President of the party since 1962, and who symbolized their link to the pre- Stroessner period.

April 1987: President Stroessner announced that the state of siege, which had been renewed every 60 days since 1956, was to be lifted, as the violence necessitating it had sufficiently abated.

February 1988: Openly identifying himself with the militant faction of the Colorado party, Stroessner was re-elected to the presidency, securing 88.6% of the votes cast, according to official results, although opposition leaders alleged serious electoral fraud. The militant faction of the Colorado party ousted *tradicionalistas* from important positions in government and increased repression against the parties of the AN. Relations with the USA deteriorated as the militants resisted attempts by the US Government to open a political dialogue with the AN.

12 January 1989: An extensive re-organization of the armed forces was engineered by the Militants to remove control of the troops from Gen. Andrés Rodríguez (the President's son-in-law) and other anti-Stroessner army officials. Rodríguez made common cause with Dr Luis María Argaña, head of the *tradicionalistas*.

2–3 February 1989: Stroessner was deposed in a military *coup d'état*. Gen. Rodríguez, who had led the rebellion, was sworn in as interim President.

May 1989: General Andrés Rodríguez was elected President, as the official candidate of the Colorado Party, receiving 74.2% of the votes cast. The parties of the AN failed to present a united candidate for the presidency, and their poor performance was widely attributed to the fragmentation of support. In concurrent elections to both chambers of Congress, the Colorado Party won 72.8% of the ballot, and secured two-thirds of the seats in each chamber. Despite Rodríguez's attempts to reunite the Colorado Party, the party effectively split in two when one faction held its own party congress and elected a list of its own supporters to a new governing council. The new Government established a commission to assess the needs of the rural population, although its impact in terms of land distribution was limited.

December 1989: The President publicly discounted any possibility of a coalition with the PRLA despite internal disagreements in the Colorado Party.

26 March 1991: The Treaty of Asunción was signed by Argentina, Brazil, Paraguay and Uruguay, committing the four countries to the establishment of a common market, to be known in Portuguese as the Mercado Comum do Sul and in Spanish as the Mercado Común del Sur (Mercosul/Mercosur—the Southern Common Market).

May 1991: Municipal elections were held in which, for the first time in Paraguayan history, mayors were directly elected by the population. The divided Colorado Party performed badly in these elections. An independent candidate, Carlos Filizzola, was elected as the capital's Mayor. He led an independent movement, Asunción para Todos, which gathered support from professional groups and the independent trade union body, CUT.

July 1991: The Federación Nacional Campesina, a national peasant association, was founded, reflecting growing social discontent despite the country's moves towards increased political democratization. The President charged his Minster of the Interior, Gen. Orlando Machuca Vargas, with the establishment of a united Traditionalist wing of the Colorado Party from among the most powerful factions.

1 December 1991: In elections to the Constituent Assembly convened to draft a new constitution, the Colorado Party received 55% of votes cast, the PLRA 27% and Constitución para Todos (CPT—a national movement which had developed from Asunción para Todos) 11%. President Rodríguez confirmed that he would not be seeking re-election upon the expiry of his mandate.

23 June 1992: The new Constitution entered into force, establishing a democratic system of government and dramatically improved protection of human rights. President Rodríguez boycotted the session of the Constituent Assembly which approved the document, amid a dispute over the provision that a serving head of state could not seek re-election (despite having announced his unwillingness to serve a further term, President Rodríguez sought to prevent the provision from affecting his current term).

December 1992: The closely contested primary election for the party's presidential candidacy was narrowly won by Argaña over Juan Carlos Wasmosy, the preferred candidate and former business associate of President Rodríguez.

March 1993: The primary result was overruled on appeal to the party's electoral tribunal, principally because of military opposition to Argaña's previous close involvement with the Stroessner regime; Wasmosy was declared the winner. He was chosen by Rodríguez despite widespread opposition by the Colorado party.

9 May 1993: Wasmosy was elected President, with 40% of the votes cast, becoming the first civilian President of Paraguay for 39 years. International observers declared the elections to have been fair.

18 August 1993: President Wasmosy appointed Gen. Lino Cesar Oviedo Silva as Commander-in-Chief of the Army; the appointment was criticized owing to Gen. Oviedo's opposition to Argaña's candidacy and his subsequent statements to the effect that an opposition victory would not be accepted. Wasmosy faced an opposition majority in both the Chamber of the Deputies and the Senate.

3 September 1993: Congress voted to repeal a law on the re-organization of the military; President Wasmosy vetoed the decision.

7 October 1993: The signing of a 'governability pact' with the opposition was suspended, following a demonstration outside the Congress building by Colorado party supporters.

14 October 1993: A preliminary agreement on a governability pact was signed by the Colorado Party and the PLRA.

29 October 1993: The President signed a new Labour Code passed by Congress; the new code had been vetoed by Rodríguez during his term of office.

29 November 1993: The Senate confirmed the approval of a law under which Congress would select Supreme Court judges.

December 1993: The President announced his decision to veto the law regarding selection of Supreme Court judges, claiming that it eroded the presidential prerogative. Congress rejected President Wasmosy's veto of proposed reforms to the Armed Forces.

9 December 1993: President Wasmosy dismissed the Minister of Foreign Affairs, Diógenes Martínez, following a disagreement regarding the administration of the Itaipú hydroelectric complex.

7 January 1994: Blas Riquelme resigned as President of the Colorado Party, following disagreements with President Wasmosy.

12 January 1994: After protracted negotiations a consensus agreement regarding the selection of Supreme Court judges, was announced.

February 1994: Serious agricultural unrest erupted in the impoverished San Pedro department and spread rapidly to other areas. The peasants demanded an equitable distribution of land and an increase in the minimum price paid for cotton.

14 February 1994: The agricultural protests escalated rapidly when police dispersed demonstrators at a roadblock with rubber bullets and tear-gas.

26 February 1994: The Central Nacional de Trabajadores (CNT) and the CUT, organized a protest march in support of the peasants' demands. The President offered concessions, including authorizing the peasants to sell their cotton directly to Brazilian buyers who were initially offering better prices than their Paraguayan intermediaries. However, these concessions failed to prevent the protests.

28 February 1994: The President agreed to hold direct discussions with the peasant leaders; the demonstrations were suspended.

4 March 1994: Protest resumed after a second series of inconclusive negotiations in which President Wasmosy refused to accept demands for a minimum price for cotton.

15 March 1994: Some 10,000 peasants marched in Asunción demanding agricultural reform following the revelation that US $85m. distributed through the National Development Bank to allow landowners to finance the purchase of cotton and soybeans had been diverted to other uses.

2 May 1994: A one-day general strike was held, Paraguay's first since 1958, with the support of the CNT and CUT. The major demands included the cessation of the Government's privatization programme, a 40% increase in wages and the suspension of Paraguay's membership of Mercosur (which was due to enter into force at the end of the year.

September 1994: President Wasmosy carried out a reshuffle of the command of the armed forces thereby strengthening the position of Gen. Oviedo. However, tensions between the two men, which had escalated since Gen. Oviedo's appointment, remained.

31 December 1994: Mercosul/Mercosur was established; its provisions came into effect the following day.

22 April 1996: Gen. Oviedo began campaigning for the leadership of the Colorado party, hoping to succeed Wasmosy as President in 1998. Wasmosy ordered Gen. Oviedo to resign; when Oviedo refused and, with the support of some 5,000 troops demanded the resignation of President Wasmosy, the latter sought refuge in the US embassy.

24 April 1996: Strengthened by regional backing and popular support, President Wasmosy agreed to a compromise whereby Gen. Oviedo would retire from active service in exchange for appointment to the Government, in the defence portfolio. Congress refused to ratify Oviedo's appointment, however.

29 April 1996: Argaña was again elected President of the Colorado Party.

13 June 1996: Gen. Oviedo was arrested and detained on a charge of sedition.

7 August 1996: The Courts of Appeal acquitted Gen. Oviedo.

September 1996: Gen. Oviedo narrowly defeated Argaña to become the Colorado Party's presidential candidate. The President attempted to exclude Oviedo from presidential contest.

October 1996: An arrest warrant was issued against Gen. Oviedo for making inflammatory public statements. Having initially avoided capture, Gen. Oviedo surrendered to the authorities in December.

March 1998: The Special Military Tribunal convened by President Wasmosy found Gen. Oviedo guilty of rebellion and sentenced him to 10 years' imprisonment.

18 April 1998: The Supreme Electoral Tribunal annulled Gen. Oviedo's presidential candidacy. The Colorado Party's new candidate was Raúl Cubas Grau, Argaña agreeing to seek election to the vice-presidency.

10 May 1998: In the presidential election, Cubas Grau gained 55.4% of the votes, defeating Laíno, of the Alianza Democrática.

15 August 1998: Cubas Grau assumed office and a new Council of Ministers, including two pro-Oviedo military officers, was inaugurated.

18 August 1998: President Cubas Grau issued a decree commuting Gen. Oviedo's period of imprisonment to that already served. The new Congress immediately rejected the decree and voted to initiate proceedings to impeach the President for unconstitutional behaviour.

2 December 1998: The Supreme Court ruled that President Cubas Grau's decree commuting Gen. Oviedo's sentence was unconstitutional.

5 December 1998: Gen. Oviedo was expelled from the party by the Argaña-supporting central committee. The motion to impeach President Cubas Grau did not receive the required two-thirds majority in congress.

14 March 1999: Argaña was expelled from the Colorado Party's headquarters by supporters of President Cubas Grau and Gen. Oviedo.

23 March 1999: Argaña was assassinated after regaining control of the Colorado Party headquarters. Argaña's supporters accused President Cubas Grau of instigating the assassination and demanded his resignation.

25 March 1999: Six protesters were killed demonstrating outside the Congress building. An official of the finance ministry was filmed firing shots at the crowd from a nearby building.

28 March 1999: President Cubas-Grau resigned and fled to Brazil where he was granted political asylum. Gen. Oviedo was subsequently granted asylum in Argentina.

29 March 1999: The President of the Senate, Luis González Macchi, was sworn in as President to serve until the end of Cubas-Grau's term in 2003. He announced the formation of a Government of National Unity the following day.

April 1999: The Supreme Court ruled that elections to the vacant vice-presidency would be held in November (the election was subsequently postponed until 2000).

May 1999: President González Macchi forcibly retired more than 100 army officers, including several high-ranking supporters of the exiled Gen. Oviedo.

12 July 1999: Gen. Oviedo was officially expelled from the Colorado Party.

15 July 1999: The arrest of Gen. Oviedo was ordered; the Argentine authorities twice refused to extradite him.

9 December 1999: Gen. Oviedo left Argentina, after indications that the country's newly elected President, Fernando de la Rúa, would treat any extradition request made by Paraguay more favourably.

February 2000: The PLRA withdrew from the Government of National Unity when it emerged that, contrary to the coalition agreement, the Colorado Party intended to present a candidate in the delayed vice-presidential election.

18 May 2000: Rebellious soldiers thought to be sympathetic to Gen. Oviedo seized a barracks. The insurrection was swiftly suppressed by the Government, which declared a 30-day national state of emergency. Some 70 people were arrested, mostly members of the security forces.

11 June 2000: Gen. Oviedo was arrested in Brazil.

13 August 2000: The PLRA Candidate, Julio César Franco, narrowly defeated the Colorado Party's candidate, Felix Argaña (son of the assassinated Vice-President) in the election to the vice-presidency.

September 2000: The election of Franco, whose candidacy was endorsed by Gen. Oviedo, caused the Minister of Defence and Public Works, José Alberto Planás, to resign. He was replaced by Walter Bower, formerly the Minister of the Interior; this portfolio was assumed by Julio César Fanego. The Minister of Finance, Federico Zayas, was replaced by Francisco Oviedo Brítez.

February 2001: The Minister of Education and Culture, Nicanor Duarte Frutos, resigned in order to campaign for the Presidency of the Colorado Party. Tensions between President González Macchi and the Franco escalated when the latter demanded the President's resignation, claiming that his status as President, despite not having been elected undermined the country's democratic status.

March 2001: The Government's problems intensified when three of the Encuentro Nacional (EN) members of the Council of Ministers threatened to resign in response to allegation that the President's car was illegally smuggled into the country. To restore confidence a cabinet reshuffle took place in which José Antonio Moreno Rufinelli was appointed to the post of Minister of Foreign Affairs, replacing Juan Esteban Aguirre.

late March 2001: Some 10,000 farmers marched through Asunción to protest against the Government's economic policy and to demand greater action on rural issues. Several thousand rural workers who had camped outside the Congress building to protest at depressed cotton prices abandoned their protest having received government concessions.

May 2001: Duarte Frutos secured the presidency of the Colorado Party with 50% of the votes cast.

8 May 2001: The President of Central Bank of Paraguay, Washington Ashwell, was forced to resign over his alleged involvement in the fraudulent transfer of US $16m. to a US bank account. In response to allegations that the President had been a beneficiary of the misappropriated funds, opposition parties submitted a congressional motion for his impeachment.

June 2001: A judge in Paraguay issued an extradition request for the country's former leader, Alfredo Stroessner, who was living in exile in Brazil. The charges concerned corruption and abuses of human rights.

Peru

c. **2500 BC:** Small villages, inhabited by farmers and fishermen, were established in the fertile river valleys on the northern coast of Peru.

c. **900 BC:** Civilizations such as the Chavín, Chimú and Nazca began migrating into the region from the north. As farming became more productive, commerce grew and states began to develop along the coast.

AD 500–1000: Large urban settlements were developed in the southern region by both the Tiahuanaco and Huari cultures.

1200–1437: The Incas of Cuzco began to expand and incorporate neighbouring civilizations.

1438–93: Under the leadership of Pachacuti Inca Yupanqui and Topa Inca Yupanqui, the Inca Empire grew to control a population of some 16m. inhabitants.

1528–32: A civil war seriously weakened the Inca Empire which, at this time, encompassed most of Ecuador, Peru, western Bolivia, northern Chile and north-western Argentina.

1531: The Spaniard, Francisco Pizarro, arrived in northern Peru with a force of 179 men.

15 November 1532: Pizarro's forces defeated the Inca army of Atahualpa at Cajamarca.

15 November 1533: Pizarro, reinforced with an army of about 5,000 men, captured the imperial city of Cuzco.

1535: The Spanish founded Ciudad de los Reyes (City of the Kings), which was subsequently renamed Lima.

1541: Pizarro was assassinated in his palace in Lima, during a conflict for power between factions of the Spanish colonists.

1544: The Viceroyalty of Peru was formally established, comprising most of southern Spanish America, with its capital in Lima.

1572: Sporadic resistance to Spanish occupation continued until the Incas were finally defeated when Túpac Amarú I, the last Inca Emperor, was captured at the fall of Vilcabamba. He was subsequently taken to Cuzco and beheaded.

1780: Túpac Amarú II led an unsuccessful rebellion against the Spanish. He was captured and executed.

1820: Gen. José de San Martín, the Argentine independence fighter, landed with his republican army in southern Peru. He had already expelled the Spanish from Chile.

28 July 1821: The republican forces of Peru formally declared the colony's independence from Spain.

1824: Independence was achieved when the republicans, assisted by the followers of the Venezuelan rebel leader, Simón Bolívar, and British soldiers, defeated Spain at the decisive battle of Ayacucho.

1826–45: Andres Santa Cruz became the first President of the Republic. He was replaced the following year by José de la Mar and in the ensuing years, rival groups competed for power and the resources of the state, several presidents, the majority of whom were military leaders, held office.

1845–51: Ramón Castilla occupied the presidency. During his time in office and a subsequent term in 1855–62, he introduced a number of reforms, including the abolition of slavery, and the construction of railway and telegraph networks. Castilla also instituted the Republic's first liberal Constitution in 1860.

1864–66: The Government entered into a war with Spain, following its seizure of the Chincha Islands. With the assistance of Bolivia, Ecuador and Chile, the Spanish were defeated.

1879–83: Peru and Bolivia were defeated by Chile in the War of the Pacific, which was fought over a territorial dispute. Under the terms of the peace agreement Chile annexed the Peruvian province of Tarapac and occupied Arica and Tacna, the latter being returned to Peruvian sovereignty in 1929.

1928: Opposition to the dictatorial regime of President Augusto Bernardino Leguía resulted in the creation of the nationalist revolutionary movement, the Alianza Popular Revolucionaria Americana (APRA).

1930: Leguía was deposed and APRA formally became a political party after its leader, Víctor Raúl de la Torre, returned from enforced exile in Mexico.

1931–45: APRA was banned by the military Government.

29 January 1942: The Rio de Janeiro Protocol was signed by Peru and Ecuador, with Argentina, Brazil, Chile and the USA as guarantors, to establish the border between the two countries. The two countries had engaged in a number of conflicts regarding the Cordillera del Cóndor; despite the agreement, confrontations occurred on a number of further occasions during the 20th century.

1948–1959: APRA was again banned by the Government.

29 October 1948: The elected President, Dr José Luis Bustamente y Rivera was deposed by Gen. Manuel Odría, following a right-wing military rebellion.

28 July 1956: Dr Manuel Prado y Ugartache succeeded Odría as President.

18 July 1962: An inconclusive presidential election precipitated military intervention, and power was assumed by Gen. Ricardo Pérez Godoy, at the head of a military junta.

3 March 1963: Pérez was replaced by his second-in-command, Gen. Nicolás Lindley López.

9 June 1963: At presidential elections Fernando Belaúnde, the joint candidate of the Acción Popular (AP) and the Partido Demócrata Cristiano, was elected President.

1965–1966: An increase in internal disturbances, attributed to Communist subversion, resulted in the suspension of constitutional guarantees and an intensive military campaign of counter-insurgency.

3 October 1968: Continuing internal unrest prompted renewed military intervention. Gen. Juan Velasco Alvarado assumed the presidency, dissolved Congress and appointed a military cabinet.

29 August 1975: President Velasco was overthrown by Gen. Francisco Morales Bermúdez.

28 July 1977: Morales announced plans for the restoration of civilian rule.

18 June 1978: A national election was held to select members of a 100-member constituent assembly, which was to draft a new constitution. APRA emerged as the largest party in the assembly.

July 1979: The new Constitution was adopted, under which the franchise was extended to the illiterate population.

1980–1985: The emergence of the Maoist terrorist group, the Sendero Luminoso (SL—Shining Path) led to the designation of an emergency zone, which extended to 13 provinces (primarily in the departments of Ayacucho, Huancavelica and Apurímac). There was a dramatic increase in violent deaths, and violations of human rights by the security forces were reported.

18 May 1980: Belaúnde decisively won the presidential election and AP achieved an outright majority in the Cámara de Diputados (Chamber of Deputies) and the greatest representation in the Senado (Senate).

July 1980: The new Constitution became fully effective and the new organs of state were inaugurated.

28 January–1 February 1981: Conflict was briefly resumed with Ecuador over the Cordillera del Cóndor.

14 April 1985: The APRA candidate, Alan García Pérez, received 45.7% of the votes cast in the first round of a presidential election, while in concurrent congressional elections APRA secured a majority in both houses of Congress.

May 1985: García's victory was assured following the withdrawal of his closest opponent, Dr Alfonso Barrantes Lingán (of the left-wing Izquierda Unida coalition) from the second round of voting.

28 July 1985: Alan García was inaugurated as President.

July 1987: The Government announced plans to nationalize banks and private financial and insurance institutions. The plans encountered considerable opposition and prompted the creation of Libertad, a 'freedom movement', which was established under the leadership of the well-known writer, Mario Vargas Llosa.

1989: Violent attacks by SL and, to a lesser extent, the Movimiento Revolucionario Túpac Amarú (MRTA) against political and industrial targets, as well as rural self-defence groups, reached an unprecedented level. Since 1980 an estimated 200,000 people had been driven from their homes and about 18,000 people killed, mostly in the department of Ayacucho.

8 April 1990: In presidential elections Vargas Llosa, the candidate of the center-right FREDEMO alliance (established in 1988 by the AP, Libertad and the Partido Popular Cristiano) obtained the largest percentage of votes against a hitherto little-known agronomist, Alberto Fujimori of the Cambio 90 group of independents.

10 June 1990: In the second round of presidential elections Fujimori won with some 57% of the votes cast.

28 July 1990: Fujimori was inaugurated and a new centre-left Council of Ministers was announced. Juan Carlos Hurtado Miller of the AP was appointed Prime Minister and Minister of the Economy and Finance.

8 August 1990: Proposals for economic readjustment were announced in which subsidies for consumers were abolished, thereby increasing prices by more than 3,000% for petrol and by about 600% for basic foodstuffs.

14 February 1991: Hurtado resigned from the Council of Ministers, following several weeks of internal dispute regarding the efficacy of anti-inflationary economic measures. Carlos Torres y Torres Lara was appointed Prime Minister.

18 June 1991: Congress granted the Government a 150-day period of emergency legislative powers in order to address the country's potentially destabilizing economic and security problems.

31 October 1991: Torres resigned as Prime Minister, in response to criticism of his handling of renewed border tension with Ecuador. He was succeeded by Alfonso de los Heros Pérez Albela.

5 April 1992: President Fujimori announced the immediate suspension of the 1979 Constitution and the dissolution of Congress, pending a comprehensive restructuring of the legislature. The 'constitutional coup' (or 'autogolpe') was implemented with the full co-operation of the armed forces and without recorded casualties. President Fujimori maintained that reform of the legislature was necessary in order to combat the armed insurrectionist movements and drugs-trafficking, and to fully implement free-market economic policies.

6 April 1992: The Prime Minister, Pérez Albela, resigned in protest against the 'autogolpe', and was replaced by Oscar de la Puente, although most government ministers elected to remain in office.

21 April 1992: The First Vice-President of the Congress, Maximo San Román Cáceres, was sworn in as the head of an alternative 'constitutional government' in front of 200 members of the deposed legislature. However, these actions were undermined by a lack of domestic and international support.

12 September 1992: The SL's leader and founder, Abimael Guzmán Reynoso, was captured. He was subsequently found guilty of treason and sentenced to life imprisonment.

22 November 1992: Elections were held to a unicameral constituent congress, the Congreso Constituyente Democrático (CCD), which was to draft a new constitution. The pro-Government parties secured 44 of the 80 congressional seats. The main opposition parties, with the exception of the Partido Popular Cristiano, boycotted the elections.

6 January 1993: The 1979 Constitution was reinstated and the CCD confirmed Fujimori as constitutional head of state.

28 August 1993: A new Council of Ministers, under the leadership of Alfonso Bustamente y Bustamente, was appointed.

31 October 1993: At a referendum the new Constitution, which, *inter alia*, introduced the death penalty for convicted terrorists and permitted a President to be re-elected for a successive five-year term of office, was approved by 52.2% of the votes cast.

29 December 1993: The new Constitution was promulgated at a ceremony boycotted by opposition members of the CCD.

February 1994: Bustamente resigned after controversial legislation was passed, which allowed a case against 11 army officers, accused of involvement in the abduction and murder of 10 alleged SL supporters in La Cantuta in 1992, to be tried in a military court. The US Government issued a formal protest at what it regarded as a breach of the autonomy of the judiciary. Nine officers received prison sentences of between one and 20 years. Efraín Goldenburg Schreiber replaced Bustamente as Prime Minister.

January 1995: Serious fighting broke out between Peru and Ecuador over the disputed region of Cordillera del Cóndor.

17 February 1995: Peru and Ecuador signed the Itamaraty Peace Declaration but hostilities began again later that month.

28 February 1995: Under the auspices of the guarantor countries of the Rio de Janeiro Protocol, a new cease-fire was agreed in Montevideo (Uruguay), between Peru and Ecuador.

9 April 1995: In presidential elections Fujimori secured an outright victory over the former UN Secretary-General, Javier Pérez de Cuéllar, obtaining 64.4% of the votes cast. Fujimori's coalition movement, Cambio 90-Nueva Mayoría (C90-NM) secured a majority in the unicameral Congress.

16 June 1995: Legislation was promulgated, which entered into immediate effect, granting amnesty to all members of the military, police and intelligence forces convicted of human-rights violations committed since 1980 in the internal conflict against armed guerrilla groups.

28 July 1995: Fujimori was inaugurated and a new Government was installed, under the leadership of Dante Córdova Blanco.

1 August 1995: An agreement between Peru and Ecuador on the extent of a demilitarized zone in the disputed Cenepa river region came into effect.

22 August 1995: An international warrant was issued for the arrest of the former President, Alan García, who had been accused of receiving bribes and other corrupt practices during his term of office.

3 April 1996: There was a cabinet reorganization following the resignation of Córdova. Alberto Pandolfi Arbúlu was appointed Prime Minister.

19 September 1996: A minor cabinet reshuffle was instigated.

29 October 1996: The foreign ministers of Peru and Ecuador signed the Santiago Agreement in Chile, which was to provide a framework for a settlement of the border issue.

17 December 1996: MRTA activists launched an assault on the residence of the Japanese ambassador in Lima, detaining by force more than 500 people.

27 December 1996: President Fujimori declared a state of emergency in Lima, effectively suspending all constitutional rights regarding arrest and the search of property.

22 April 1997: Peruvian troops launched an assault on the Japanese ambassador's residence, ending the MRTA siege and killing all 14 of the MRTA activists.

16 July 1997: A minor cabinet reshuffle was implemented.

5 June 1998: Javier Valle Riestra replaced Pandolfi as Prime Minister in a cabinet reorganization.

7 August 1998: Pandolfi was reappointed Prime Minister.

26 October 1998: An accord was signed in the Brazilian capital, Brasília, between Peru and Ecuador, confirming Peru's claim regarding the delineation of the border, but granting Ecuador navigation and trading rights on the Amazon and its tributaries, in addition to a small enclave of territory within Peru where a number of soldiers killed in battle were buried.

5 January 1999: As part of a major cabinet reorganization, Víctor Joy Way Rojas replaced Pandolfi as Prime Minister.

15 April 1999: Five ministers were replaced amid internal government divisions caused by corruption allegations.

14 July 1999: Oscar Ramírez Durand, the leader of Sendero Rojo (a dissident SL faction), was captured. He was subsequently sentenced to life imprisonment.

13 October 1999: A new cabinet was appointed under the leadership of Alberto Bustamente Belaúnde.

1 January 2000: The Jurado Nacional Electoral (JNE) ratified Fujimori's eligibility to seek election to a third presidential term, on behalf of the Perú 2000 alliance, on the grounds that he had only served one term under the current Constitution and was, therefore, eligible for re-election (the Constitution prevented any citizen from seeking election to a third presidential term).

9 April 2000: At the first round of the presidential election, Fujimori officially secured 49.9% of the votes cast; Alejandro Toledo, of Perú Posible (PP), received 40.2% of the ballot. In congressional elections Perú 2000 won 52 of the 120 seats while PP won 29.

May 2000: Toledo withdrew from the second round of the presidential election as disputes persisted regarding the legitimacy of the electoral process.

28 May 2000: Despite the suspension by the Organization of American States (OAS) and other organizations of the programme of election-monitoring, the second round proceeded with Fujimori effectively unopposed (although Toledo remained a candidate because the JNE refused to remove his name from the ballot papers). Fujimori won 74.3% of the valid votes cast.

28 July 2000: Fujimori's inauguration was accompanied by violent protests in Lima.

September 2000: A major political scandal erupted after the disclosure of a video recording that allegedly showed Vladimiro Montesinos, head of the national intelligence service, bribing an opposition member of Congress.

16 September 2000: President Fujimori announced that new elections, in which he would not participate, would be held.

5 October 2000: Congress approved OAS-mediated proposals, which facilitated the transfer of power from President Fujimori to his successor in mid-2001.

23 October 2000: The first Vice-President, Francisco Tudela, resigned.

16 November 2000: Congress voted to replace its pro-Fujimori President, Martha Hildebrandt Pérez Treviño, with Valentín Paniagua.

20 November 2000: Shortly after arriving in Japan, Fujimori resigned the presidency. He was subsequently granted Japanese citizenship, thus effectively preventing his extradition.

21 November 2000: Congress refused to accept Fujimori's resignation and instead voted to dismiss him from the presidency.

22 November 2000: Paniagua was appointed interim President.

25 November 2000: A new cabinet was sworn in, under the leadership of Pérez de Cuellar.

9 January 2001: A Truth Commission was established to investigate the fate of an estimated 4,000 people who had disappeared during the campaign against the guerrilla groups in the 1980s and 1990s.

23 February 2001: Fujimori was charged with dereliction of duty.

8 April 2001: In presidential elections Toledo won 36.5% of the vote, Alan García (who had returned to Peru in January after corruption charges against him were ruled to have expired under the statute of limitations) won 25.8% of the ballot and Lourdes Flores of the Unidad Nacional (UN) won 24.3%. In congressional elections no party secured a majority in the 120-seat Congress: PP gained 45 seats, APRA 27 and UN 17.

3 June 2001: Toledo was elected President with 53.1% of the second-round votes cast, compared with García's 46.9%.

23 June 2001: Venezuelan military-intelligence officers arrested Montesinos in that country's capital, Caracas.

28 July 2001: Toledo was inaugurated as the first Peruvian President of indigenous descent. A new cabinet, led by Roberto Danino, took office.

30 October 2001: Congress voted to revoke Fujimori's immunity from prosecution and charges of embezzlement and illicit enrichment were filed against him.

Saint Christopher and Nevis

c. **AD 400:** The Arawak people settled in the Caribbean area.

c. **1400:** The Kalinago tribe, known as the Caribs, settled in the Caribbean area, ousting the Arawaks.

1493: The Italian navigator in the service of the Spanish Crown, Cristoforo Colombo (Christopher Columbus) sighted the island, naming it after his patron saint, St Christopher.

28 January 1624: The British, led by Sir Thomas Warner, colonized the island, abbreviating the name to St Kitts (although the island remains officially known as St Christopher).

1626: British and French settlers killed the entire Carib population at Bloody Point, on the west coast of St Christopher.

1628: The British colonized the neighbouring island of Nevis.

1664: The French gained complete control of St Christopher.

1689: The British regained control of St Christopher.

1706: France re-captured St Chrisopher.

1713: St Christopher was ceded to the British through the Treaty of Utrecht.

1782: The French besieged the island, resulting in British surrender after nine months.

1783: St Christopher was again ceded to the British, through the Treaty of Paris.

1816: Saint Christopher-Nevis-Anguilla and the British Virgin Islands were administered as a single colony.

1871: The Leeward Islands Federation was formed.

1880: The British placed St Christopher-Nevis-Anguilla under a single administration.

30 June 1956: The Leeward Islands Federation was dissolved and Antigua and Barbuda, Saint Kitts-Nevis-Anguilla, Montserrat and the British Virgin Islands became British colonies, allowing them to enter the proposed Caribbean Federation as independent units.

2 August 1956: The British Parliament adopted a bill establishing the British Caribbean Federation, which was to have a joint Government, Legislature and Supreme Court.

1958: St Christopher-Nevis-Anguilla joined the West Indies Federation.

6 October 1959: The British Colonial Office announced the abolition of the post of Governor of the Leeward Islands as part of the reform of the constitutions of the Leeward Islands. The reforms were designed to reduce the administrative differences between the members of the West Indies Federation.

27 February–5 March 1962: Representatives of Barbados and the seven Leeward and Windward Island colonies met in Barbados. They requested that the British Government approve a new federation, known as the Little Eight, in the event of the collapse of the West Indies Federation.

31 May 1962: The British Government formally dissolved the West Indies Federation, which had collapsed.

17 November 1962: General elections were held in St Christopher-Nevis-Anguilla. The ruling Labour Party retained power and the party's leader, Paul Southwell, formed a new Government.

18 April–4 May 1966: A Constitutional Conference was held in London to discuss the future of Dominica, Grenada, St Christopher-Nevis-Anguilla, St Lucia and St Vincent following the collapse of the proposed Little Seven federation (as it had become following Grenada's withdrawal).

1 November 1966: Antigua, Barbados, Dominica, Grenada, Montserrat, St Christopher-Nevis-Anguilla, St Lucia and St Vincent founded the West Indies (Associated States) Council of Ministers, which discussed proposals for a 'small island federation'. These talks, however, collapsed and the islands, with the exception of Montserrat, subsequently achieved full independence separately.

27 February 1967: St Christopher-Nevis-Anguilla was granted the status of an Associated State, and became self-governing, although the United Kingdom retained responsibility for defence and foreign relations. The Legislative Council was replaced by a House of Assembly, the Administrator became Governor, and the Chief Minister, Robert Bradshaw, leader of the Labour Party, became Premier.

19 March 1967: A force of 200 soldiers belonging to the British Parachute Regiment occupied Anguilla and installed Anthony Lee as Administrator.

31 May 1967: Around 250 armed Anguillans forced the island's 17 policemen to leave, thereby achieving virtual independence.

11 July 1967: A referendum on Anguilla overwhelmingly approved independence for the island, and Peter Adams took office as President.

4 August 1967: Adams was deposed as President of Anguilla and replaced by the Rev. Ronald Webster, following the approval by Adams of an unpopular compromise agreement with the Government of St Christopher and Nevis.

10 May 1971: Bradshaw retained his position as Prime Minister, after the Labour Party won seven of the nine seats in the House of Assembly. The two seats for Nevis were secured by the People's Action Movement (PAM) and the Nevis Reformation Party (NRP).

25 July 1972: The Governments of Guyana, Dominica, Grenada, St Christopher-Nevis-Anguilla, St Lucia and St Vincent signed the Declaration of Grenada, which created the framework for an Eastern Caribbean political union.

30 May 1975: The British Government announced that Anguilla would be granted separate status within the existing Associated State.

1 December 1975: The Labour Party was returned to office at a general elections and Bradshaw was reappointed Premier.

10 February 1976: The new Constitution proposed by the British Government came into effect in Anguilla and St Christopher-Nevis-Anguilla, and elections were subsequently held to Anguilla's separate House of Assembly.

23 May 1978: The Deputy Premier, Tim Southwell, took office as Premier following the death of Bradshaw.

18 May 1979: Southwell died and was succeeded as Premier by the former Attorney-General and Minister of External Affairs, Lee Moore.

18 February 1980: The Labour Party was defeated at general elections by a coalition of the People's Action Movement (PAM) and the Nevis Reformation Party (NRP). The PAM leader, Dr Kennedy A. Simmonds, became Premier.

19 December 1980: Anguilla formally withdrew from the Associated State of St Christopher-Nevis-Anguilla, which had not exercised any control over the island's affairs since 1971, and became a separate dependency of the United Kingdom.

19 September 1983: The islands achieved full independence from the United Kingdom as Saint Christopher and Nevis. A new Constitution was introduced and Simmonds became Prime Minister.

21 June 1984: The PAM-NRP coalition was returned to power at legislative elections, winning nine of the 11 seats in the newly expanded National Assembly.

21 March 1989: The PAM won six of the 11 seats at elections to the National Assembly and Simmonds was reappointed Prime Minister.

29 November 1993: At a general election neither the PAM nor the Labour Party managed to secure a majority in the House of Assembly, with both parties winning four seats. On Nevis the Concerned Citizens' Movement (CCM) secured two seats and the NRP one. The leader of the CCM, Vance Amory, refused to form a coalition with either PAM or the Labour Party, obliging the Governor-General to invite Simmonds to form a minority Government with the support of the NRP.

1–2 December 1993: Violent demonstrations were held in Basseterre to protest against Simmonds' attempts to form a new Government, and a state of emergency was declared.

3 July 1995: The instability of the Government's majority forced Simmonds to call a general election, at which he was conclusively defeated. The Labour Party won a large majority and the party's leader, Denzil Douglas, was appointed Prime Minister.

23 June 1996: Vance Amory, the Premier of Nevis, announced he was to launch a bid for Nevis to secede from the federation with St Christopher.

24 February 1997: The CCM won three of the five elected seats in the Nevis Island Assembly, while the NRP won the remaining two.

13 October 1997: The five elected members of the Nevis Island Assembly voted to secede from the federation and announced a referendum on the subject.

10 August 1998: The referendum on the secession of Nevis was approved by around 62% of voters, although this fell short of the two-thirds majority required for a change of status. Douglas subsequently pledged greater autonomy for the island.

February 2000: A report published into allegations of corruption within the Simmonds Government, dating back to 1997, made 10 specific accusations of negligence, improper behaviour and irresponsible action against Simmonds. However, the accusations were rejected by Simmonds.

6 March 2000: The Labour Party retained power at legislative elections, taking all eight seats on St Christopher and 64.5% of votes cast; the PAM obtained 35.5%. The distribution of seats on Nevis remained unchanged. Douglas was reappointed Prime Minister. The PAM subsequently accused the Government of electoral fraud and announced that the party would not contest any future elections until the electoral register had been revised.

26–27 June 2000: The Organisation for Economic Co-operation and Development (OECD) and the Financial Action Task Force (FATF, part of the OECD) criticized St Christopher and Nevis' financial system, naming it as a tax haven. The OECD requested that the country reform its rules by 31 December 2005 or face sanctions.

8 October 2000: Following the resignation of Simmonds as leader of the PAM, Lindsay Grant was elected in his place.

Saint Lucia

***c.* AD 400:** The Arawak people settled in the Caribbean area.

***c.* 1400:** The Kalinago tribe, known as the Caribs, settled in the Caribbean area, ousting the Arawaks. They named the island Hewanorra.

***c.* 1500:** The Spanish explorer, Juan de la Cosa, was believed to be the first European to land on the island, naming it St Lucia.

1600: The Dutch settled on St Lucia, establishing Vieux Fort.

1605: The British landed on St Lucia and settlers remained there. Following attacks by Carib inhabitants, however, they left the island.

1639: A second attempt by the British to colonize St Lucia also failed.

1651: Representatives of the French West India Company purchased the island. However the sale was subsequently disputed by British merchants.

1746: The town of Soufrière was founded by the French, who established sugar plantations and imported slave labourers.

1814: Saint Lucia was ceded to the United Kingdom, which declared it a Crown Colony.

1838: Slavery was abolished, resulting in the subsequent decline of the sugar industry.

1924: St Lucia was granted representative government.

1951: At general elections, held, for the first time under universal adult suffrage, the Saint Lucia Labour Party (SLP) was victorious and Georges Charles was appointed Chief Minister.

2 August 1956: The British Parliament adopted a bill establishing the British Caribbean Federation, which was to have a joint Government, Legislature and Supreme Court.

1958: St Lucia joined the West Indies Federation.

6 October 1959: The British Colonial Office announced the abolition of the post of Governor of the Windward Islands as part of the reform of the constitutions of the Leeward Islands. The reforms were designed to reduce the administrative differences between the members of the West Indies Federation.

14 April 1961: The SLP won nine of the ten available seats in the Legislative Council and Charles remained Chief Minister.

27 February–5 March 1962: Representatives of Barbados and the seven Leeward and Windward Island colonies met in Barbados. They requested that the British Government approve a new federation, known as the Little Eight, in the event of the collapse of the West Indies Federation.

31 May 1962: The British Government formally dissolved the West Indies Federation, which had collapsed.

1964: John (later Sir John) Compton of the United Workers' Party (UWP) became Chief Minister.

1 November 1966: Antigua, Barbados, Dominica, Grenada, Montserrat, St Christopher-Nevis-Anguilla, St Lucia and St Vincent founded the West Indies (Associated States) Council of Ministers, which discussed proposals for a 'small island federation'. These talks, however, collapsed and the islands, with the exception of Montserrat, subsequently achieved full independence separately.

1 March 1967: St Lucia was granted the status of an Associated State, and became self-governing, although the United Kingdom retained responsibility for defence and foreign relations. The Legislative Council was replaced by a House of Assembly, the Administrator was designated Governor and the Chief Minister became Premier.

25 July 1972: The Governments of Guyana, Dominica, Grenada, St Kitts and Nevis, St Lucia and St Vincent signed the Declaration of Grenada, which created the framework for an Eastern Caribbean political union.

6 May 1974: The UWP won a majority in the expanded House of Assembly, and Compton was appointed Premier. The party had fought the election on its policy of independence for the island, and subsequently began negotiations on the subject with the British Government.

22 February 1979: St Lucia achieved full independence from the United Kingdom, with Compton as Prime Minister.

2 July 1979: At a general election, the SLP defeated the UWP. The SLP leader, Allan Louisy, was appointed Prime Minister.

30 April 1981: Louisy resigned as Prime Minister following disagreements with a separate faction of the SLP, led by George Odlum. Winston Cenac replaced him as Prime Minister, and Odlum subsequently founded the Progressive Labour Party (PLP).

16 January 1982: Cenac resigned in the wake of massive public protests against his administration's economic policies. The deputy leader of the PLP, Michael Pilgrim, was appointed interim Prime Minister.

3 May 1982: The UWP secured a substantial victory at legislative elections, and John Compton again became Prime Minister.

6 April 1987: The UWP secured a majority of one at elections to the House of Assembly. Compton initially decided to form a Government but subsequently chose to call new elections with the aim of achieving a larger majority.

30 April 1987: The UWP failed to improve on its previous tally of seats, and Compton was reappointed Prime Minister.

June 1987: The UWP's majority in the House of Assembly increased to three seats, following Cenac's defection from the SLP.

7 October 1988: The Government assumed control of the banana industry until June 1990, despite protests against its actions.

27 April 1992: The UWP increased its majority at legislative elections, winning 56.3% of the votes cast and securing 11 seats; the SLP took the remaining six seats. Compton was reappointed as Prime Minister.

October 1993: a three-day strike was organized by the Banana Salvation Committee (BSC), in support of demands for an increase in the minimum price paid to local producers for bananas; two people were killed in the protests. Price increases were later implemented by the Government.

March 1994: A strike was organized by the BSC in protest at the Government's decision to place the Saint Lucia Banana Growers' Association (SLBGA) into receivership.

February 1996: Industrial action was undertaken by the BSC in protest at the monopoly on banana exports exercised by the state-run Windward Islands Banana Development and Exporting Co (Wibdeco).

2 April 1996: Following Compton's resignation, Vaughan Lewis was appointed Prime Minister.

1 June 1996: The SLP criticized the appointment of George Mallet as Governor-General on the grounds that Mallet, a former deputy leader of the UWP, would politicize the post.

September 1996: Legislation was approved effectively transferring the control of the SLBGA from the Government to the banana growers.

October 1996: Further industrial action by banana growers took place in support of demands to end Wibdeco's monopoly. Violent clashes ensued between banana growers and farmers in opposition to the strike.

23 May 1997: The SLP secured a landslide victory at a general election, winning 16 of the 17 seats in the House of Assembly and obtaining 61.3% of the votes cast; the UWP obtained 36.6% of the ballot. The leader of the SLP, Kenny Anthony, became Prime Minister.

17 September 1997: Perlette Louisy was appointed Governor-General in place of Mallet, who had resigned in August following pressure from the SLP.

July 1998: The SLBGA was privatized and the Saint Lucia Banana Corpn (SLBC) was created in its place.

October 1998: A court judgment ruled that the SLBC did not enjoy monopoly rights in the banana trade; the ruling followed complaints from the Tropical Quality Fruit Co (TQF) that Wibdeco had been refusing to accept its fruit for shipment, allegedly as a result of pressure exerted by the SLBC Chairman, a member of the Wibdeco board.

26–27 June 2000: The Organisation for Economic Co-operation and Development (OECD) and the Financial Action Task Force (FATF, part of the OECD) criticized St Lucia's financial system. The OECD urged the Government to reform its rules by 31 December 2005 or face penalties.

October 2000: Lewis announced his resignation as leader of the UWP. Morella Joseph, one of the party's two deputy leaders was subsequently elected to replace him.

29 March 2001: The Minister of Foreign Affairs and International Trade, George Odlum, resigned after publicly stating he would not contest the next election as an SLP candidate. Odlum joined a new opposition group, the National Alliance (NA), that had been founded by Compton and Joseph, in advance of legislative elections, scheduled for December 2001.

Saint Vincent and the Grenadines

5000 BC: The islands now known as St Vincent and the Grenadines were inhabited by the Ciboney people.

***c.* AD 400:** The Arawak people settled in the Caribbean area.

***c.* 1400:** The Kalinago tribe, known as the Caribs, settled in the Caribbean area, ousting the Arawaks.

1498: The Italian navigator in the service of the Spanish Crown, Cristoforo Colombo (Christopher Columbus), first saw the island, naming it St Vincent.

1675: A Dutch slave ship, stranded on the island, brought African slaves who were allowed to settle on the island, alongside the Carib inhabitants.

1700: Tension between African slaves and the Caribs led to civil war.

***c.* 1700:** French settlers arrived on St Vincent.

1722: British settlers arrived on St Vincent, and possession of the island was disputed.

1763: St Vincent was ceded to British control.

1778: The French regained control of the island.

1783: St Vincent reverted to British rule under the Treaty of Paris.

1834: Slavery was abolished.

1951: Universal adult suffrage was introduced.

2 August 1956: The British Parliament adopted a bill establishing the British Caribbean Federation, which was to have a joint Government, Legislature and Supreme Court.

1958: St Vincent joined the West Indies Federation.

6 October 1959: The British Colonial Office announced the abolition of the post of Governor of the Windward Islands as part of the reform of the constitutions of the Leeward Islands. The reforms were designed to reduce the administrative differences between the members of the West Indies Federation.

20 April 1961: At elections to the Legislative Council, the ruling People's Political Party (PPP) won six of the nine seats available and Ebenezer Joshua remained as Chief Minister.

27 February–5 March 1962: Representatives of Barbados and the seven Leeward and Windward Island colonies met in Barbados. They requested that the British Government approve a new federation, known as the Little Eight, in the event of the collapse of the West Indies Federation.

31 May 1962: The British Government formally dissolved the West Indies Federation, which had collapsed.

1966: The island's transfer to Associated Status was delayed after a disagreement developed over electoral boundaries following a disputed general election won by the PPP. The British Government negotiated an agreement by which new boundaries were to be established and new elections held.

1 November 1966: Antigua, Barbados, Dominica, Grenada, Montserrat, St Christopher-Nevis-Anguilla, St Lucia and St Vincent founded the West Indies (Associated States) Council of Ministers, which discussed proposals for a 'small island federation'. These talks, however, collapsed and the islands, with the exception of Montserrat, subsequently achieved full independence separately.

May 1967: The leader of the St Vincent Labour Party (SVLP), Milton Cato, was appointed Chief Minister after winning a general election held under the new boundary system.

27 October 1969: St Vincent and the Grenadines was granted the status of an Associated State, and became self-governing, although the United Kingdom retained responsibility for defence and foreign relations.

7 April 1972: The SVLP and PPP both won six seats at a general election. James Mitchell, an independent, secured the remaining seat. Mitchell agreed to form a Government with the PPP and he was appointed Premier, with Joshua as Deputy Premier and Minister of Finance.

25 July 1972: The Governments of Guyana, Dominica, Grenada, St Christopher-Nevis-Anguilla, St Lucia and St Vincent signed the Declaration of Grenada, which created the framework for an Eastern Caribbean political union.

18 September 1974: The House of Assembly passed a vote of 'no confidence' in the Government after Joshua and his wife resigned from the Government following a disagreement with Mitchell over policy issues.

9 December 1974: The SVLP won 10 of the 13 seats in the House of Assembly and the SVLP leader, Milton Cato, returned to office as Premier.

April 1978: Cato dismissed Joshua as Minister of Trade and Agriculture, effectively ending the coalition between the SVLP and the PPP.

27 October 1979: The islands attained full independence from the United Kingdom under the name of St Vincent and the Grenadines. Cato became Prime Minister and Sir Sydney Gun-Munro became Governor-General.

5 December 1979: The SVLP maintained its majority at a general election, obtaining 11 of the 13 elective seats in the 19-member House of Assembly, and Cato was reappointed Prime Minister.

7 December 1979: A group of rebels, led by Lennox 'Rasta Bomba' Charles, captured civil and police installations on Union Island in protest at the return to power of the Cato Government, which they accused of neglecting the Grenadines. Troops from Barbados subsequently ended the revolt.

25 July 1984: The incumbent SVLP Government was defeated by the centrist New Democratic Party (NDP) at legislative elections. The NDP won nine of the 13 elective seats in the House of Assembly and the party's leader, James Mitchell, became Prime Minister.

22 January 1985: Cato retired from politics and was replaced as SVLP leader by Hudson Tannis. Vincent Beache took over as leader of the opposition in the House of Assembly.

21 February 1985: Joseph Eustace replaced Gun-Munro as Governor-General.

3 August 1986: Hudson Tannis was killed in an airplane crash and was replaced as SVLP leader by Vincent Beache.

16 May 1989: The NDP won all 15 elective seats in the House of Assembly at a general election, and Mitchell retained his position as Prime Minister.

24 January 1994: The SVLP and the Movement for National Unity (MNU) formed an electoral alliance, which later became formalized as the United Labour Party (ULP), led by Beache.

21 February 1994: The NDP was returned to power with a reduced majority at legislative elections, obtaining 12 seats in the House of Assembly, and Mitchell was reappointed Prime Minister. The ULP alliance won the three remaining elective seats.

September 1995: Parnell R. Campbell, the former Deputy Prime Minister, Attorney-General and Minister of Justice, resigned, following allegations of financial impropriety. Campbell, however, denied any wrongdoing.

29 August 1996: The Attorney General and Minister of Justice, Carlyle Dougan, resigned after he was implicated in a fraudulent financial transaction.

15 June 1998: The NDP won a fourth successive victory at legislative elections to the House of Assembly, obtaining eight seats; the ULP secured the remaining seven seats. Mitchell retained his position as Prime Minister. The ULP, which had won 54.6% of votes cast, called on the NDP to hold new elections within nine months; furthermore, the ULP made allegations of electoral irregularities, which were strongly denied by Mitchell.

9 July 1998: At the opening session of the House of Assembly, several hundred protesters demonstrated against the new Government, and two ministers were attacked.

December 1998: Beache resigned as leader of the ULP, and Dr Ralph Gonsalves was elected in his place. Beache remained leader of the opposition in the House of Assembly.

January 2000: The ULP threatened to propose a motion of 'no confidence' in the Government, in protest at government plans to increase the salaries of members of the House of Assembly. The ULP accused the Government of corruption, of excessive expenditure and of neglecting rising crime levels and the education sector.

April 2000: The Government introduced legislation increasing the salaries and benefits of members of the House of Assembly, provoking demonstrations and industrial action. The protesters, grouped under the Organization in Defence of Democracy (ODD) demanded the resignation of the Government and new elections.

4 May 2000: The Government and the opposition reached an accord, which aimed to end the recent political protests, whereby new legislative elections would be called before March 2001.

26–27 June 2000: The Organisation for Economic Co-operation and Development (OECD) and the Financial Action Task Force (FATF, part of the OECD) criticized St Vincent and the Grenadines' financial system. The OECD urged the Government to reform its rules by 31 December 2005 or face penalties.

20 August 2000: Mitchell resigned as NDP leader and was replaced by the Minister of Finance, Arnhim Eustace.

27 October 2000: Mitchell resigned as Prime Minister in favour of Eustace, under the terms of an agreement with the opposition designed to end the anti-Government protests that had affected the islands since May.

28 March 2001: The ULP defeated the incumbent NDP at legislative elections, winning 12 of the 15 parliamentary seats and 57% of votes cast. The NDP secured 40.9% of the votes and the remaining three seats in the House of Assembly. The ULP leader, Ralph Gonsalves, was appointed Prime Minister.

26 April 2001: Gonsalves reorganized the Offshore Finance Authority in order to increase the regulation of the offshore financial sector.

Suriname

***c.* 3000 BC:** Indigenous tribes, the most important of which was the Arawaks, settled in the area now constituting Suriname.

15th century AD: The Surina tribe, the original inhabitants of what became Suriname, were driven out of the territory by other Amerindian groups.

1499: The northern coast of South America was discovered by the Italian navigator, Amerigo Vespucci.

late 16th century: The Dutch commenced the establishment of trading posts along the Commewijn and Corantijn rivers (both in modern-day Suriname).

1593: Spain officially claimed possession of the region.

1621: The Dutch Westindische Compagnie (West India Company—WIC) was formed to exploit the region.

1650: The English colonial Governor of Barbados, Sir Francis Willoughby, successfully colonized the region, establishing tobacco and sugar plantations on the west bank of the Suriname river, and founded the settlement now known as Paramaribo (now the capital of present-day Suriname).

1667: The Dutch, under the leadership of Abraham Crijnsen, invaded and conquered the English settlement. The territory was ceded to the Netherlands under the Treaty of Breda, in exchange for the Dutch colony of Nieuw Amsterdam (now New York, USA), and named Dutch Guiana.

1674: An uprising of the indigenous population and the Maroons was violently suppressed by the Dutch military.

late 17th century: Slaves (who became known as Maroons or Bush Negroes—Boschnegers) were imported to the territory from Africa by the Dutch to work on the plantations.

1683: The territory was transferred to the WIC, the city of Amsterdam and the van Sommelsdijk family, each of which owned one-third thereof. Cornelis van Aerssen heer van Sommelsdijk became the first Governor of Dutch Guiana.

1688: Van Sommelsdijk was murdered by a group of discontented soldiers during a mutiny, although order was swiftly restored in the territory by the new Governor, Gen. van Scherpenhuijsen.

early 18th century: The Maroons began to flee the colonial plantations in large numbers and formed settlements in the interior, from which they launched numerous attacks on the tobacco and sugar plantations.

1749: Governor Johan Jacob Mauricius signed a peace treaty with the Maroons of the Saramaccan tribe.

1761: Governor Wigbold Crommelin signed a further peace treaty with the Maroons, which guaranteed their liberty and granted them an annual supply of weapons and munitions. In return the Maroons pledged to return all future runaway slaves to the plantations and never to appear in Paramaribo in armed groups of more than six persons.

late 18th century: Further slave revolts occurred in the territory.

August 1799: During the Napoleonic Wars, English forces invaded and occupied Dutch Guiana.

25 March 1802: The Treaty of Amiens ended the War of the Second Coalition and returned the territory to the French-created Batavian Republic in the Netherlands; however, in May 1804 the British recaptured Dutch Guiana.

1808: The trading of slaves in the territory was prohibited by the British occupying regime.

February 1816: Following the Congress of Vienna (1814–15) and the subsequent signing of the Treaty of Paris in November 1815, which ended the European wars with revolutionary France (the Napoleonic Wars), Dutch Guiana was formally returned to the newly created Kingdom of the United Netherlands. The colony was henceforth administered by a Governor, with the assistance of a Political Council, the members of which were appointed by the Governor, following nomination by the colonial planter-class.

1828: The administration of all Dutch West Indies colonies was centralized under a Governor-General (the first being Johannes Graaf van den Bosch), stationed in Dutch Guiana, who reported directly to the Colonial Office in the Netherlands.

1 July 1863: King William III of the Netherlands abolished slavery in all the Dutch colonies. The freed slaves were, nevertheless, forced to work at the plantations for a further 10 years, although they received wages for doing so.

1866: The Koloniale Staten (Colonial Assembly), a representative body with limited local power, was established. However, ultimate power continued to reside in The Hague.

1922: The colony was designated an integral part of the Netherlands.

1949: Universal adult suffrage was introduced.

1950: Suriname was granted internal self-government by the Dutch.

15 December 1954: Suriname became an autonomous region within the Kingdom of the Netherlands, which also included the Netherlands Antilles. The Dutch retained the right to rule in matters of defence and foreign affairs.

1960s: The Nationale Partij Suriname (NPS), led by Johan Pengel and the Verenigde Hindostaanse Partij (VHP), led by Jaggernath Lachmon, formed a Creole-Hindustani alliance.

1969: Pengel was replaced as Prime Minister by Dr Jules Sedney of the Progressieve Nationale Partij (PNP).

1970: Pengel died, he was replaced as leader of the PNP by Henck Arron.

November 1973: An alliance of the NPS, the Partij Nationalistische Republiek (PNR) gained 22 of the 39 seats in the legislature. Arron was appointed Prime Minister.

25 November 1975: Full and complete independence was granted to Suriname by the Dutch. Johan Ferrier became the first President.

25 February 1980: The elected Government was overthrown by 16 non-commissioned army officers, led by Sgt-Maj. Desi Bouterse. Dr Hendrik Chin-A-Sen was appointed Prime Minister of a new Government, which was named the Nationale Militaire Raad (NMR).

August 1980: In a further *coup d'état*, Ferrier was dismissed and Chin-A-Sen was appointed President, retaining the office of Prime Minister. The legislature was dissolved and a state of emergency was declared.

February 1982: Bouterse and the NMR overthrew Chin-A-Sen's Government and appointed L. Fred Ramdat Misier as President. A state of siege and martial law were subsequently imposed.

December 1982: Members of the armed forces burned down the offices of the opposition; several prominent citizens were killed in the ensuing unrest, an event which became known as the 'December Murders'. The Government was subsequently dissolved and an interim military government appointed. The Netherlands and the USA halted all deliveries of aid to the region.

February 1983: Bouterse formed a civil-military Council of Ministers, with Dr Errol Alibux as Prime Minister.

August 1984: The state of emergency was lifted.

January 1985: A nominated National Assembly, which comprised representatives of Bouterse's 25th November Movement, trade unions and the business sector, was inaugurated.

November 1985: Arron, Lachmon and members of the Vooruitstrevende Hervormings Partij (VHP) and the Kaum-Tani Persuatan Indonesia (KTPI), joined the NMR, which had been renamed the Topberaad (Supreme Council).

31 March 1987: A draft for a new constitution was approved by the National Assembly.

September 1987: In a referendum, 93% of voters endorsed the new Constitution.

November 1987: In a general election to the National Assembly, the Front voor Demokratie en Ontwikkeling (FDO) won a decisive victory.

January 1988: The National Assembly elected Ramsewak Shankar as President and Henck Arron as Vice-President.

24 December 1990: Following the resignation of Bouterse as Commander-in-Chief of the armed forces, a military *coup d'état* took place. Johan Kraog of the NPS was appointed Provisional President; he invited Bouterse to resume the office of Commander-in-Chief.

7 January 1991: A transitional Government was inaugurated.

25 May 1991: In general elections, the Nieuw Front (NF), which comprised members of the former FDO, the Surinaamse Partij van de Arvid was victorious, gaining 30 seats in the National Assembly.

7 September 1991: Runaldo R. Venetiaan of the NF was elected President.

March 1992: Amendments to the Constitution were approved, which provided for the reduction of the armed forces by two-thirds, and the defence budget by 50%. Measures to curb the political influence of the military were also introduced.

1995: The Government and Guyana signed an agreement on the establishment of a joint commission to resolve the territorial dispute over the Corentje region.

23 May 1996: Following a general election in which no single party gained an overall majority, Jules Wijdenbosch was elected President by the Vereinigde Volksvergardering (United People's Assembly), a body consisting of members of the National Assembly and district and municipal councils.

August 1997: The Government recalled its ambassador to the Netherlands, following its request for an international arrest warrant for Bouterse, on charges of illegal drugs-trafficking.

June 1998: A general strike took place, the protests demanded the resignation of the Government in favour of a non-political administration.

March 1999: The trial of Bouterse *in absentia* by the Dutch authorities began, the following July he was sentenced to 16 years' imprisonment (later reduced to 11 years) and fined US $2.3m.

May 1999: Following a further general strike, the entire Cabinet resigned; the President refused to accept the resignations, however, and the Government remained in office.

1 June 1999: The National Assembly adopted a motion of 'no confidence' in President Wijdenbosch by 27 votes to 14 (with 10 abstentions), thus failing to obtain the two-thirds majority required in order to oblige the President to resign. President Wijdenbosch remained in office, although he made a commitment to organize a general election for May 2000.

December 1999: Following the implication of many of its members in financial and sexual scandals, the Government resubmitted its resignation, which was accepted.

May 2000: The Government claimed that Guyana had violated its territory by granting the Canadian company, CGX Energy Inc, permission to explore disputed maritime waters for petroleum. Following the intervention of Surinamese troops, the company was forced to withdraw in July.

25 May 2000: At a general election, the NF secured 33 of the 51 seats in the National Assembly, the Millenium Combinatie 10 and the DNP 2000 three.

August 2000: Venetiaan was elected President.

1 November 2000: The Suriname Court of Justice ruled that Bouterse would be compelled to stand trial for his role in the 'December Murders' of 1982. Several days later, the Dutch High Court ruled that Bouterse would be tried *in absentia* for torture and other crimes against humanity.

20 March 2001: Fred Derby, a survivor of the 'December Murders' of 1982, died. He had been an important witness in the trials of Bouterse, in both Suriname and the Netherlands.

Trinidad and Tobago

AD 1498: The Italian navigator in the service of the Spanish Crown, Cristoforo Colombo (Christopher Columbus) claimed the island now known as Trinidad for Spain. At that time the island was inhabited by Amerindian and Carib groups. The Amerindians were all gradually eliminated. Columbus also named Bella Forma (later re-named Tobago, a corruption of tobacco).

1532: Trinidad was colonized by the Spanish.

1592: The Spanish established settlements, including San José de Oruña (later St Joseph), which formed part of the Viceroyalty of New Granada, based in Bogotá (Santafé de Bogotá, Colombia).

1628: The Dutch colonized Tobago.

1641: King Charles I of England granted Tobago to the Duke of Courland, and a British settlement was established on the northern side of the island.

1658: The Dutch regained control of Tobago.

1662: A French settler, Cornelius Lampsius, claimed Tobago for France.

1762: The British took possession of Tobago, and the island was ceded to them in the following year, under the Treaty of Paris. However, control of the island remained in dispute between the British and the French.

1797: Trinidad was captured by British forces.

1802: Trinidad was formally ceded to the United Kingdom by Spain, under the Treaty of Amiens. Africans were transported to the island to work as slaves on the plantations.

1814: The island of Tobago was again ceded to the United Kingdom.

1834: Slavery was abolished, followed by a four-year obligatory 'apprenticeship' system.

1877: Tobago became a British Crown Colony.

1889: The islands of Trinidad and Tobago were combined to form one British colony.

1899: The colonial authorities abolished the elected Borough Council in the capital, Port of Spain.

1914: The Borough Council was re-established.

1925: The Legislative Council became partly elected, although the electoral roll remained limited.

19 January 1949: The British Colonial Secretary, Creech Jones, published details of a new Constitution for Trinidad and Tobago, which was intended to increase residents' involvement in government.

18 September 1950: The first elections were held to the expanded Legislative Council. No party obtained an absolute majority; Uriah Butler's 'Home Rule' Party received the most votes.

26 May 1956: The Legislative Council was dissolved after approving a five-year development programme costing around £18m.

2 August 1956: The British Parliament adopted a bill establishing the British Caribbean Federation, which was to have a joint Government, Legislature and Supreme Court.

24 September 1956: The People's National Movement (PNM), founded by Dr Eric Williams, secured a majority to the new 24-member Legislative Council. Williams was later appointed the islands' first Chief Minister.

1958: The territory became a member of the West Indies Federation.

20 July 1959: A new Constitution came into force, which introduced a ministerial system of government. Williams became Premier and appointed ministers to a new cabinet.

28 May–8 June 1962: The British Government agreed to grant Trinidad and Tobago full independence at a constitutional conference held in the United Kingdom.

31 May 1962: The British Government formally dissolved the West Indies Federation, which had collapsed partly as a result of the withdrawal of Jamaica and of Trinidad and Tobago.

31 August 1962: The islands achieved full independence from the United Kingdom.

4 December 1962: Elections were held to the new 30-member House of Representatives. The PNM won 20 seats and Williams formed a new Government.

7 November 1966: The PNM won its third consecutive victory at legislative elections and Williams remained as Prime Minister.

1 May 1968: Trinidad and Tobago became a member of the Caribbean Free Trade Association (CARIFTA), the establishment of which had been agreed by Antigua, Barbados and British Guiana in December 1965.

21 April 1970: The Government declared a state of emergency following violent demonstrations by 'Black Power' supporters, who demanded improved economic opportunities for the island's black community. The Trinidad and Tobago Regiment (the country's army) mutinied within hours of the announcement.

24 April 1970: The army mutiny collapsed and the rebel officers and 'Black Power' leaders were arrested and charged with treason and sedition.

24 May 1971: The PNM won all the seats in the House of Representatives at elections boycotted by the main opposition parties, which had formed an Action Committee for Dedicated Citizens (ACDC) to protest at the Government's organization of the elections.

1 August 1976: Trinidad and Tobago adopted a republican Constitution, and the former Governor-General, Sir Ellis Clarke was sworn in as interim President.

13 September 1976: The PNM won 53% of the vote at a general election and Williams was reappointed Prime Minister.

29 December 1976: Clarke was formally elected President by the country's legislature.

21 April 1980: The PNM won a resounding victory at local elections in Trinidad. Elections in Tobago were postponed due to a lack of agreement over their format.

24 November 1980: At local elections in Tobago, the Democratic Action Congress (DAC) won eight of the 12 seats in the new Tobago Assembly.

29 March 1981: Williams died suddenly and George Chambers, the Minister of Agriculture, Lands and Fisheries, and Industry and Commerce, became acting Prime Minister.

9 May 1981: Chambers was confirmed as Prime Minister at a special conference of the ruling PNP.

9 November 1982: At legislative elections, the PNM was returned to power for its sixth successive term, and Chambers was reappointed Prime Minister.

August 1983: Opposition candidates won 66 of the 120 local government seats contested at nationwide elections after the four major opposition parties successfully co-ordinated their electoral strategy.

16 November 1984: The Democratic Action Congress (DAC) won 11 of the 12 seats in the Tobago House of Assembly after campaigning against the PNM's alleged failure to improve conditions on the island.

15 December 1986: The opposition grouping, the National Alliance for Reconstruction (NAR), defeated the PNM at legislative elections. The NAR leader, Arthur Robinson, was appointed Prime Minister.

19 March 1987: Clarke retired as President and Noor Hassanali was elected in his place by legislators.

15 May 1988: Four former NAR ministers were expelled from the party for challenging Robinson's leadership. They subsequently announced their intention of forming a new party.

28 November 1988: At elections to the Tobago House of Assembly, the distribution of seats remained unchanged.

27 July–1 August 1990: A militant black Muslim group, Jamaat al-Muslimeen, launched an armed uprising against the Government. At least 30 people were killed before the security forces re-established order. The uprising led to the postponement of local elections scheduled for September.

16 December 1991: The PNM comprehensively defeated the NAR in a general election held amid growing popular concern about the country's poor economic position. The PNM leader, Patrick Manning, was appointed Prime Minister, while Robinson resigned from the leadership of the NAR.

19 May 1995: Manning carried out a major cabinet reshuffle.

20 June 1995: The former Attorney General and Minister for National Security, Selwyn Richardson, was assassinated.

3 August 1995: The Government declared a state of emergency in order to contain possible protests during the removal of the Speaker of the House of Representatives, Occah Seapaul, who had been accused of perjury. The Minister of Public Utilities, Seapaul's brother, Ralph Maraj, resigned in protest at the proceedings.

6 November 1995: Mannin called an early election after his majority was reduced to one. At the elections, the PNM and the United National Congress (UNC), led by Basdeo Panday, both won 17 seats, while the NAR won two. Basdeo Panday subsequently established a coalition with the NAR and was appointed Prime Minister.

5 December 1995: Panday met the leader of the Jamaat al-Muslimeen to discuss their demands for land.

14 February 1997: The former Prime Minister, Arthur Robinson, was elected President, despite claims by the PNM that Robinson would favour the NAR.

4–7 June 1999: Nine members of a gang of drug dealers, who had been found guilty of multiple murder, were executed. The hangings, the first since 1994, attracted widespread international condemnation.

31 December 1999: A prominent UNC member, Hansraj Sumairsingh, was assassinated, allegedly after threatening to expose corruption in the administration of the Government's unemployment relief programme.

12 October 2000: The Minister for Local Government, Dhanraj Singh, was dismissed after investigations began into his role in the assassination of Sumairsingh.

11 December 2000: Panday was reappointed Prime Minister after he led the UNC to victory at a general election. The UNC secured 19 of the 36 seats in the House of Representatives, while the PNM won 16 seats and the NAR gained one.

14 February 2001: President Robinson finally agreed to approve the appointment as senators of seven UNC members who had been defeated as candidates to the House of Representatives. Robinson had initially refused to approve the appointments, which he described as a 'rebuff to the electorate'.

18 May 2001: The High Court awarded TT $625,000 in damages to the Jamaat al-Muslimeen, to compensate the group for the destruction of its headquarters by the security forces.

United States of America

c. **30,000 BC:** During the late Pleistocene Epoch of the Quaternary Period, the last Ice Age exposed a land bridge, subsequently submerged beneath the Bering Sea, across which Asian peoples migrated from the north-eastern part of what is now Siberia to what is now Alaska. They gradually spread throughout the Americas and were the first known inhabitants of the continent. Sparse archaeological evidence indicates that these peoples were probably nomadic hunter-gatherers who lived in groups of about 100 individuals, used stone tools and tanned reindeer or caribou skins.

11,000–8500 BC: Archaeological discoveries near the modern town of Clovis, in eastern New Mexico, reveal that a civilization there used spear points and stone blades of a style not found in other regions of the world.

8000 BC: By this time, climate change had caused polar ice to melt and raise the sea-level above the Bering Sea land-bridge, ending overland migration into the Americas from Asia.

6000 BC–5000 BC: Archaeological evidence suggests the presence of distinct cultures in areas now in California and Texas during this period.

2000 BC: Finds in Georgia and Florida indicate that indigenous Americans of the Southeast made pottery by this time.

1000 BC: The so-called 'Mound Builders' flourished in the valleys of the Mississippi and Ohio Rivers; they were the ancestors of the people living in this area more than 2,000 years later when the first Europeans arrived there. The earliest mounds, known as midden mounds, were filled with domestic refuse, suggesting that by this time the people who built them lived in permanent villages and practised settled agriculture, cultivating crops including maize and beans.

200 BC–AD 300: Mound City flourished in the south of the modern state of Ohio; archaeological evidence suggests that the various indigenous groups of the region had interacted with each other, exchanged technologies and traded extensively by this time.

c. **700:** The Hohokam migrated north from modern Mexico into the south-western area of the USA. They developed the technology of irrigation, harnessing water from the Salt and Gila Rivers to grow maize, beans, marrows and avocados.

***c.* 1025:** The Navajo migrated southward into what would become Arizona, Colorado, New Mexico and Utah.

1050–1250: The Pueblo built cities of interconnected terraces on cliff-faces in central Arizona; 22 such cities survive.

1100–1300: The mound-building tradition, shared by large numbers of tribes and tribal groupings, reached its zenith at this time.

1490: When the Europeans began to arrive in North America, the population of the area now comprising the USA was about 6m. people. They comprised Eastern Woodland peoples, who inhabited the north-eastern quarter of the future USA, farmed maize and beans, and comprised 35 tribes including the Iroquois Confederacy of New York, the Winnebago of Wisconsin, and the Peoria and Illinwa of Illinois; the Southeast peoples, who lived in an area extending from eastern Texas to Maryland to the southern tip of Florida, traded extensively, developed large towns, grew maize and comprised 48 tribes, the best known of which were the Cherokee, the Chickasaw, the Choctaw, the Creek and the Seminole, whom the Europeans dubbed the 'Five Civilized Tribes'; the Great Plains peoples, who inhabited the prairies from Texas to the Dakotas, were nomadic and comprised 27 tribes including the Cheyenne, the Arapaho, the Sioux and the Comanche; the Southwest peoples, who lived in and around Arizona and New Mexico, farmed maize and other crops by means of irrigation, built adobe cliff dwellings and cities and comprised 27 tribes including the Navajo and the Hopi; the Plateau peoples, comprising 22 tribes in and around inland Washington, Oregon and Idaho; the North-West Pacific Coast peoples, 6 tribes, including the Chinook, who inhabited a thin strip all along the Pacific coast of Washington and Oregon; and the California Intermountain peoples, 47 tribes who lived primarily in California and Nevada.

12 October 1492: The Genovese Italian navigator Cristoforo Colombo (known in English as Christopher Columbus), sailing in the service of the Spanish joint monarchs King Fernando of Aragon and Queen Isabel of Castile, landed on an island now considered to be Guanahaní (Samana Cay), in the Bahamas group. Columbus erroneously believed that he had arrived at an island near India, his destination. For this reason, he called the local people of this island 'Indians'; this name for indigenous Americans was retained after the discovery that this was a new continent. Columbus was not the first European to visit the New World (Leif Ericsson had founded a temporary settlement on Newfoundland in Canada in 1000), nor did Columbus realize at the time that he had discovered a new continent. Nevertheless, it is his voyage credited with establishing the American continent as a destination for European explorers and colonists.

19 November 1493: During his second voyage of discovery, Columbus landed on the island of Boriquen, now known as Puerto Rico and an 'autonomous commonwealth' of the USA. Columbus named the island San Juan de Bautista, which persists in the name of Puerto Rico's capital city, San Juan.

1497–98: The Venetian Italian navigator Giovanni Caboto, sailing under the name John Cabot in the service of King Henry VII of England, explored along North America's east coast between Labrador, in what is now Canada, and Chesapeake Bay, in what would become the states of Virginia and Maryland. Cabot may have been the first European to visit the future USA.

1497–98: The Florentine Italian navigator Amerigo Vespucci, by now a resident in the Spanish city of Seville, stated that during a first voyage he had made at this time, he had become the first European to reach the North American mainland. Despite doubts about the validity of his claim, the continent was subsequently known by Latinized form of his forename.

2 April 1513: Juan Ponce de León, searching for the legendary 'Fountain of Youth' about which he had heard the native Borinqueno people of Puerto Rico speak, landed on the North American mainland, naming the area 'Florida', meaning 'flowery'.

1521: Ponce de León led some 200 mainly Spanish settlers in an unsuccessful attempt to found a colony on the west coast of Florida. Native American resistance was strong, and Ponce de León himself was fatally wounded. The party retreated to Cuba, where he died shortly after arriving.

1524: The Florentine Italian navigator Giovanni da Verrazano, sailing in the service of King François I of France, became the first European of record to visit the settlement which eventually developed into New York City.

1526: Spanish settlers made an unsuccessful attempt to found a colony in future South Carolina, at a site on the estuary of the Pee Dee River.

28 September 1542: The Portuguese explorer and soldier Juan Rodríquez Cabrillo sailed into San Diego Bay, believed to be the first instance of European landing on North America from the Pacific.

***c.* 1550:** Five Eastern Woodland tribes of what is now the northern part of New York state—the Cayuga, the Mohawk, the Oneida, the Onondaga and the Seneca—formed a confederacy called the Five Nations of Iroquois. They traded among themselves and with others using *wampum* (shell currency) and grew maize, beans, pumpkins and tobacco, not only for subsistence but as cash crops. The tribes' matrilinear system vested power in councils rather than in 'chiefs', a term applied by Europeans seeking a focal point for their dealings with the tribes. The Five Nations proved themselves to be astute at dealing with the new demands placed upon them by the arrival of the Europeans; maintaining their independence amid the rival British and French territorial ambitions.

1562: French Protestants, known as Huguenots, attempted to found a colony on Parris Island, in future South Carolina, fleeing persecution in Roman Catholic France, but were unable to remain. Another Huguenot group established a colony in Florida in 1564.

1565: The Spanish explorer Pedro Menéndez de Avilés led a party of Spanish Roman Catholics who founded the first permanent European settlement on the mainland of the future USA, at St Augustine, in north-eastern Florida. He also massacred all the settlers of the nearby French Huguenot colony of Fort Caroline.

1566: The Spanish built a fort on Parris Island; this, too, would be temporary.

1579: The English navigator and soldier Sir Francis Drake sailed up the west coast of the Americas as far north as Vancouver Island. He and his party were the first Europeans known to have seen the west coast of the Americas north of San Diego Bay. He gave this coast of America the name New Albion.

1580–1606: Chief Wahunsonacoh, better known as Powhatan, the name of the tribe of whom he was chief, formed the Powhatan Confederacy, grouping 32 Algonquin-speaking Eastern Woodland tribes on and near the west coast of Chesapeake Bay into an alliance. As their paramount chief, Powhatan would face the challenge of the arrival of European settlers in the confederacy's midst.

September 1584: The English explorer and writer Sir Walter Raleigh explored the Outer Banks of North Carolina, made contact with the indigenous peoples there, chose a site for possible colonization and returned to England, taking with him two local people, who were presented at court.

1585: Raleigh led 108 settlers in seven ships in the first attempt to found an English colony in the Americas, on Roanoke Island, in the area he had explored the previous year.

1586: Drake attacked and burned the Spanish colony of St Augustine but did not succeed in destroying it. On his voyage home his ships carried the returning Roanoke colonists, who had abandoned their settlement.

23 July 1587: Raleigh again attempted a settlement on Roanoke Island, with 121 colonists led by Governor John White. This colony also failed; an expedition sent to England for supplies returned in 1590 but found the settlement abandoned with no evidence of the colonists they had left behind.

14 May 1607: Captain Christopher Newport, in the service of the London Company during the reign of King James I of England, arrived with a group of 140 English settlers on three ships at an island in south-eastern Virginia. There they founded Jamestown, named in the King's honour; this was the first permanent English settlement in the Americas.

1614: Dutch entrepreneurs founded a trading post on the right (west) bank of the Hudson River, in the territory of the Mahicans, at the place where the city of Albany now stands. This trading station, known as Fort Nassau, was the first Dutch presence in the future USA. Although true colonization would come a decade later, Fort Nassau marked the beginning of New Netherlands, which would become New York state.

1616: An epidemic of smallpox—unknown in the Americas prior to the arrival of Europeans—killed more than one-half the indigenous population of the area corresponding to the six modern states of New England.

30 July 1619: The first European representative assembly in American history, the House of Burgesses, met in James City, as Jamestown had been renamed; 22 delegates attended, representing 11 constituencies in the Virginia colony.

August 1619: A Dutch warship, having defeated a Spanish ship in a confrontation on the high seas and seized its cargo, delivered to the James City colony some 20 African labourers. This is the first documented arrival of Africans into the future USA. There is no evidence that these earliest African-Americans were held in slavery, although they may, in common with many European immigrants, have been indentured servants.

16 September 1620: The ship *Mayflower* sailed with the tide from the English port of Plymouth, carrying 102 men, women and children, some but not all of whom were Puritan Christians escaping persecution in England, bound for the Virginia colony at Jamestown, where they had permission to settle.

November 1620: After a storm had blown the ship off course, the *Mayflower* landed in Cape Cod Bay, in what is now Massachusetts, some 800 km north of Jamestown. A party went ashore to find a suitable location for a colony, which they named Plymouth.

21 November 1620: The adult male passengers of the *Mayflower* were required to sign an agreement, the Mayflower Compact, in which they pledged to 'covenant and combine ourselves together into a civil body politik, for our better ordering and preservation' and to 'enact, constitute and frame such just and equal laws, ordinances, acts, constitutions and offices, from time to time, as shall be thought meet and convenient for the general good of the Colony'. This was the first constitution ever written in the 'New World'.

early 1621: The Plymouth colonists signed a peace treaty and shared knowledge with the Wampanoag people.

1621: The States-General in the Netherlands granted a charter to the Dutch West India Company, with a monopoly over Dutch trade with and colonization in the Americas and Africa.

late 1621: Although the exact date is not known, the Plymouth colonists held a feast of Thanksgiving for the harvest they had just gathered, inviting their Wampanoag neighbours; the festival remains celebrated in the USA.

1624: Dutch and Walloon settlers founded a permanent colony, Fort Orange, at the site of the Fort Nassau trading post. Dutch colonization of New Netherlands began. In the same year, Dutch entrepreneurs founded the New Amsterdam trading post on the southern tip of Manhattan Island.

1626: The German-born Dutch Director-General of the New Netherlands colony, Peter Minuit, purchased Manhattan Island from an Algonquin-speaking tribe.

1630s: Further groups of British colonists arrived, founding settlements along the eastern American coast, some proceeding to the interior to found further colonies.

12 June 1630: John Winthrop arrived at what is now Salem, Massachusetts, with some 1,800 mostly-wealthy Puritans to found the Massachusetts Bay Colony, first at Salem, then briefly at Charleston and then shortly thereafter at Boston.

28 October 1636: Harvard College, now Harvard University, was founded in Cambridge, Massachusetts.

1638: The Dutch West India Company, and especially Minuit, persuaded the Swedish Government to recruit Swedish and Finnish settlers to establish a permanent settlement; Fort Christina, named after the Swedish monarch, would be the beginning of the colony of New Sweden, which subsequently became the colony and state of Delaware.

1639: The Connecticut colony adopted the Fundamental Orders, which abolished religious affiliation as a requirement for citizenship.

1640: Three black indentured servants in the Virginia colony escaped, were captured and were tried. One of them, John Punch, received a sentence of servitude for life. This is the first documented case of full slavery in the English colonies of North America. In the same year, another black resident of the Virginia colony, Anthony Johnson, was a free man and a landholder.

1643: The first confederation of the American colonies was a defensive association of Connecticut, Massachusetts Bay and New Plymouth against the Dutch, the French and certain indigenous groups with which these had formed alliances.

1661: The Virginia House of Burgesses enacted the so-called 'Black Codes', which institutionalized slavery in the colony.

1664: In the Second Anglo-Dutch War, British forces seized the Dutch colony of New Amsterdam. although the Dutch successfully recaptured it shortly afterwards.

1667: The Treaty of Breda ended the Second Anglo-Dutch War and confirmed the acquisition of New Netherlands by the British; both the city of New Amsterdam and the colony of New Netherlands were renamed New York, and Fort Orange was renamed Albany. Although the number of Dutch people arriving in the New York colony declined sharply, the Dutch language and way of life endured. The first US President ever to be born within the independent USA, Martin Van Buren (who served from 1837–41), was a native speaker of Dutch.

1670: With slavery laws removing the rights of black people in the Virginia colony, and similar legislation having been enacted in other colonies, Johnson had his lands confiscated.

1671: Denmark captured St Thomas Island, in the archipelago now known as the US Virgin Islands.

1675: The Wampanoag, frustrated by the English settlers' failure to honour treaty obligations, abrogated the treaties, formed alliances with neighbouring peoples and began an uprising against the colonists.

August 1676: The Wampanoag Chief, Metacomat, died in battle; the indigenous groups of New England never again constituted a formidable threat to English expansion and usurpation of their lands.

1679: The separate royal colony of New Hampshire was created from the royal colony of Massachusetts.

1681: The British monarch, King Charles II, granted a large tract of land to William Penn, in payment of a £16,000 debt owed to Penn's late father. Penn the Younger suggested the name 'Sylvania', meaning 'forest', for the tract, but King Charles added 'Penn' at the beginning; the territory eventually became the colony of Pennsylvania.

1682: The French explorer Robert Cavelier Sieur de La Salle navigated the entire length of the Mississippi River, and claimed it and all the lands it drained for France. He named the territory Louisiana, after King Louis XIV of France.

1689–97: King William's War, named after Britain's Dutch-born King William III, was the name given to the North American theatre of the War of the Grand Alliance (also known as the War of the League of Augsburg) in Europe, in which Britain, Austria, the Netherlands and Spain formed the 'grand alliance' to oppose France. War erupted in America after a series of English raids on French settlements in Canada, when the Governor of New France retaliated by attacking and burning several English colonial towns and threatening Boston and New York City. French pirates based in Nova Scotia disrupted British ships approaching American ports. English colonial troops then captured the French colony of Port Royal in Nova Scotia.

20 September 1697: The Treaty of Ryswick concluded the War of the Grand Alliance; it returned Port Royal to France.

1702–13: Queen Anne's War was the name given to the North American theatre of the War of the Spanish Succession (1701–14) in Europe. The conflict brought British forces into opposition with both French and Spanish forces in North America.

1705: French Huguenots from Virginia established the first permanent European settlement in the future state of North Carolina.

1709–11: The Great Palatine Migration of 1710 brought an estimated 50,000 German settlers into the British American colonies. Recruited from the Palatinate area of Germany by British entrepreneurs, the majority settled in the Hudson River Valley of the New York colony, but many also settled in New Jersey and other colonies.

1710: As Queen Anne's War continued, British colonial troops, aided by the British Royal Navy, occupied the French colony of Acadia (now in Canada), although their assaults on Québec City and Montréal failed.

1712: The Carolina patent, granting control over the colony, was divided into two separate British royal colonies, which became North Carolina and South Carolina.

1713: The Treaty of Utrecht concluded the War of the Spanish Succession and Queen Anne's War; it left many issues in North America unresolved.

1715–22: During this period, the Five Nations of Iroquois became the Six Nations, when the linguistically affiliated Tuscarora, having been expelled from their territory in North Carolina by encroaching European settlers, migrated to New York state and were invited to join the confederacy.

1718: The Montréal-born French-Canadian explorer Jean-Baptiste Le Moyne Sieur de Bienville founded the city of New Orleans, as a purpose-built capital city for the vast French-held territory known as Louisiana.

12 February 1733: The British philanthropist and prison reformer James Edward Oglethorpe led 116 colonists to found the first permanent European settlement in the future state of Georgia, as a debtors' asylum, at Savannah. He would name the colony after the German-born Hanoverian King George II of England, who had granted Oglethorpe permission for the settlement.

29 July 1741: The Danish navigator Vitus Jonassen Bering, sailing in the service of the Russian Imperial Navy, entered the Gulf of Alaska and sighted the North American continent. The sea between North America and Siberian Russia was subsequently named in his honour.

1742: Oglethorpe's forces defeated the expansionist Spanish colonists of Florida, ending the Spanish threat to Georgia.

1744–48: King George's War was the name given to the North American theatre of the War of the Austrian Succession in Europe. French forces based in Canada made several attempts against Boston and New York but did not succeed in capturing them.

1754–63: The French and Indian War was the name given to the North American theatre of the Seven Years' War (1756—63) in Europe, in which Britain, the Kingdom of Prussia and the Duchy of Hannover defeated France, Austria, Russia, Sweden, Spain and the Kingdom of Saxony. There was also an Asian theatre, where the fighting in India was known as the Third Carnatic War. Britain gained military control over most of French Canada.

10 February 1763: The Treaty of Paris, which formalized the conclusion of the end of the Seven Years' War ceded most of the French Empire in North America to British control, including all of France's Louisiana territory that lay east of the Mississippi River. However, that part of Louisiana lying west of the Mississippi was not included because France had ceded it to Spain the previous year in a secret treaty. Spain ceded Florida to Britain.

1765: The British Parliament, in which the American colonies were not represented, and without consultation with the colonies, approved the Stamp Act, requiring all legal documents in Great Britain and its colonies to bear a government stamp, for which it was necessary to pay a fee. The phrase 'no taxation without representation' subsequently became popular among American colonists seeking greater autonomy.

1767: The Townshend Acts enacted by the British Parliament levied heavy taxes on a number of goods imported into Britain's American colonies. The colonists vehemently protested against them.

1769–82: The Spanish Roman Catholic Franciscan missionary Fra Junípero Serra founded a series of missions along the coast of Upper California. Starting with the first, established at San Diego in 1769, he spaced them at intervals of one day's walk. After his death, his successors would continue the effort until a chain of 21 missions linked the California coast all the way north to Sonoma, founded in 1823.

5 March 1770: In the Boston Massacre, colonialists hurled missiles at British troops who had arrived to enforce the Townshend Acts. The troops retaliated by firing their muskets into the crowd, killing five colonial men. The British commanding officer and eight British soldiers were subsequently tried for murder; two were convicted of manslaughter, the others were acquitted. The outcome caused widespread colonial resentment.

November 1773: Three ships in the service of the British East India Company sailed into Boston harbour bearing cargoes of tea from the Far East. Although the British Government had by now relaxed most of the stringent provisions of the Townshend Acts, the tax on tea remained. Massachusetts colonists refused to allow the unloading of the three ships, which remained laden in the harbour.

16 December 1773: Massachusetts citizens boarded the three ships in Boston harbour and threw their cargoes of tea overboard, in an action subsequently known to history as the Boston Tea Party. This incident came to symbolize colonial opposition to British dominance.

March 1774: The British Parliament enacted the so-called 'Coercive Acts', punitive legislation against the Massachusetts colony in retaliation over the Boston Tea Party incident. These measures, dubbed in the colonies the 'Intolerable Acts', included the Massachusetts Government Act, which revoked the colony's charter and banned town meetings; the Quartering Act, which required colonial residents to quarter and provision British troops; the Impartial Administration of Justice Act, which granted British officials immunity from prosecution in Massachusetts courts; and the Boston Port Act, which closed the port of Boston to trade and removed the seat of the colonial government from Boston to Salem.

5 September 1774: The First Continental Congress convened in Philadelphia, Pennsylvania. Although the Coercive Acts had been aimed specifically at Massachusetts, the 50 delegates represented 12 of the 14 British colonies, the absentees being Georgia and Canada. The congress drafted a Declaration of Rights and Grievances, addressed to Britain's King George III. The Congress also ratified the Continental Association, which demanded a ban on all trade with British merchants. The Congress established Committees of Safety to enforce its proclamations; these rapidly, though, they became local representations of the Continental Congress and conduits through which the colonies prepared for war. One of the responsibilities of the Committees of Safety was to arrange for the quartering and provisioning of a Continental Army.

26 October 1774: After establishing a date of 10 May 1775 by which King George III was required to respond to the Declaration of Rights and Grievances, the First Continental Congress adjourned. However, the activities of the Committees of Safety continued. By the specified date the colonies had assembled a force of 'Minutemen' (so named owing to their alleged readiness to fight at 60 seconds' notice), and were already at war with Britain.

18 April 1775: Paul Revere and two companions rode through the night from Boston to warn the Minutemen of Concord, Massachusetts that British troops were marching in their direction (the British knew of large stores of ammunition and supplies established by the Committees of Safety there). The British detained Revere *en route*, although one of his companions did reach his destination. The event became a further popular symbol of resistance to British rule.

19 April 1775: At Lexington, Massachusetts, 70 'Minutemen' under the command of Capt. John Parker confronted some 700 Concord-bound British troops under the command of Lt-Col Francis Smith. The British demanded that the Minutemen disperse, but they refused. The British then fired and battle ensued, and the Minutemen retreated. The British force marched on to Concord. The Battle of Lexington marked the beginning of the conflict known internationally as the War of American Independence, more commonly in the USA as the American Revolution or the Revolutionary War.

19 April 1775: A force estimated at 200 Minutemen succeeded in driving the British force back from Concord to Lexington and eventually to Charlestown.

10 May 1775: Having failed to receive a satisfactory response from King George III, the Second Continental Congress convened in Philadelphia; it would remain in session for nearly six years. Georgia attended on this occasion, bringing the number of colonies represented to 13. The Congress soon found itself to be effectively governing a new country, without having had executive authority officially vested in it. The Congress appointed one of the delegates, Gen. George Washington of the Virginia colony, to assemble a continental army to deal with the tense situation in Boston, where the British army lay under siege primarily by local civilians. It dispatched diplomats to Europe to negotiate alliances with other powers and issued paper money to maintain economic activity amid British embargoes, although most of the delegates considered these to be interim measures, until the signing of an agreement with the British forces.

10 May 1775: A militia from Vermont, commanded by Ethan Allen, and a Connecticut militia commanded by Col Benedict Arnold, captured the British Fort Ticonderoga, on Lake Champlain. They confiscated valuable cannons and dragged them overland to Boston.

17 June 1775: The British won a battle for control of the hills above Boston at Bunker Hill, although they suffered severe losses.

23 August 1775: King George III issued a royal proclamation stating that the colonies of New England were in a state of 'rebellion' and establishing measures for 'suppressing rebellion and sedition' in the American colonies. He subsequently hired some 20,000 Hessian German troops to strengthen the British military presence in the American colonies.

10 January 1776: The British political activist and writer, Thomas Paine, who had emigrated to Philadelphia in 1774, anonymously published a 50-page pamphlet entitled *Common Sense*. He argued persuasively that it was to the colonists' great advantage to pursue a course of full independence from Britain. The pamphlet proved highly popular and influenced colonial opinion in favour of full independence.

May 1776: The Second Continental Congress voted to instruct each colony to establish its own self-governing institutions, independent of the existing British colonial structure. Debate centred on full independence, placing every delegate in danger of treason charges against the British Crown.

2 June 1776: The Second Continental Congress approved a resolution demanding full independence for the United Colonies. Thomas Jefferson, a delegate from Virginia, was enlisted to draft a fuller version that would explain why independence was being declared.

28 June 1776: Jefferson submitted his draft to the Congress; the chamber edited the document, most notably expunging Jefferson's criticism of Britain for perpetuating the slave trade.

4 July 1776: The Second Continental Congress adopted the 'Declaration of Independence by the Congress of the United States of America'. Its adoption would eventually be unanimous.

25 December 1776: Gen. Washington's troops captured over 900 Hessian troops at Trenton, New Jersey; by early January 1777, he had gained control over the entirety of the New Jersey colony.

1777: American forces won decisive battles at Bennington and at Saratoga, New York. The British forces were weakened in number and in logistical support, and the colonial forces were dominant.

15 November 1777: After much deliberation, the Second Continental Congress approved the second draft of the 'Articles of Confederation and Perpetual Union'. The articles then required the ratification of each of the 13 individual colonies before they would come into force, a process that would take more than three years. The document provided for a weak central government, with executive powers largely vested in the 13 individual state governments.

February 1778: France formally recognized the independence of the USA; in April it sent a naval force to assist the colonial cause.

September 1779: By now Spain had also expressed its support for the Americans, and French and Spanish naval forces bombarded the British fortress at Gibraltar, opening a European theatre of the War of American Independence.

1 March 1781: The Articles of Confederation were ratified by the last remaining state, Maryland, and entered into effect. The Second Continental Congress was dissolved and was superseded by the Congress of the Confederation.

19 October 1781: The Commander of the British forces, Gen. Charles Cornwallis, surrendered to Gen. Washington at Yorktown, Virginia; hostilities ceased in the American theatre of the War of American Independence.

1782: King Kamehameha I, known as Kamehameha 'the Great', came to the throne in Hawaii and would reign until his death. His dynasty ruled the islands until its overthrow in 1893. He united the islands and succeeded in protecting their independence from domination by foreign powers in an age of European colonial expansion.

1783: The Massachusetts Supreme Court prohibited slavery in the colony.

3 September 1783: The Treaty of Paris followed the conclusion of the War of American Independence. Under this treaty, Britain formally recognized the independence of the USA and ceded to the USA all British territorial claims south of Canada, north of Mexico and east of the Mississippi River, except for Florida, which was returned to Spain.

25 May–27 September 1787: The Constitutional Convention met at the State House in Philadelphia to revise the Articles of Confederation, by now widely thought to be too weak to serve their purpose adequately. At this convention, the 55 delegates representing 12 of the 13 former colonies (Rhode Island opposed revising the articles and declined to send any delegates) drafted a new Constitution of the United States.

17 September 1787: The Constitutional Convention approved the new Constitution when 39 of the 55 delegates voted in favour of it. The document stipulated that at least two-thirds of the states—i.e. initially nine of the 13—would be required to ratify its text before it could take effect. The document provided for a presidential system of government at federal level, with legislative power vested in a bicameral legislature, known as Congress, the two chambers being named the Senate (upper house) and the House of Representatives (lower house).

7 December 1787: Delaware's legislature became the first to ratify the US Constitution; Delaware would henceforth call itself 'the First State'.

December 1787–May 1788: During this period, Pennsylvania, New Jersey, Georgia, Connecticut, Massachusetts (which demanded the creation of a written bill of rights), Maryland and South Carolina ratified the Constitution. North Carolina opposed ratification.

21 June 1788: New Hampshire became the ninth of the 13 former colonies to ratify the Constitution, thereby meeting the requirement that two-thirds of the former colonies must ratify it. The Constitution, therefore, entered into force.

26 June–26 July 1788: Virginia and New York ratified the Constitution.

1789: New York City became the first federal capital of the USA and would remain so until the following year.

March 1789: Washington, stood unopposed as a Federalist for the office of President and was unanimously elected by the Electoral College. John Adams, a Federalist from Massachusetts, was elected the first Vice-President.

4 March 1789: The first session of Congress convened in New York City. On the agenda were more than 100 proposed amendments to the Constitution. Congress reduced their number to 12; of these, two were rejected. Debate on the remaining 10 amendments began.

30 April 1789: George Washington took the oath of office on the balcony of Federal Hall in New York City and became the first President of the USA.

September 1789: The US Supreme Court was created.

21 November 1789: North Carolina, after having voted against ratification in 1788, now became the 12th state to ratify the Constitution.

1790: Philadelphia became the second federal capital of the USA.

29 May 1790: Rhode Island, after the other states placed significant pressure on it and even threatened to regard it as a 'foreign power' subject to trade tariffs and duties if it refused to ratify the Constitution, acquiesced and became the final state to do so. The unity of the original 13 colonies was preserved.

4 March 1791: Vermont, after having unilaterally declared itself to be an independent republic in 1777, joined the Union and became the 14th state of the USA.

15 December 1791: The first 10 amendments to the Constitution completed the necessary approval process and became the US Bill of Rights. The First Amendment guarantees freedom of speech, freedom of the press, the right to petition and the separation of church and state; the Second grants the people the right to bear arms; the Third protects the populace from any obligation to quarter soldiers during peacetime; the Fourth limits the right of search and seizure; the Fifth concerns criminal prosecutions; it safeguards against the seizure of property without just compensation, establishes that once acquitted of a crime, a defendant cannot then be tried again for the same offence, prohibits detention without charge, and decrees that a person may not be compelled to testify against himself; the Sixth guarantees a right to a 'speedy' and public trial; the Seventh grants the right to trial by a jury in matters where the value in controversy exceeds $20; the Eighth protects against excessive bail, excessive fines and 'cruel and unusual' punishment; the Ninth Amendment governs the rule of construction of the Constitution itself and the Tenth reserves for the states all powers not expressly granted to the federal government by the Constitution.

1792: President Washington stood unopposed for re-election and won the votes of all 132 members of the Electoral College.

1 June 1792: Kentucky, formerly part of Virginia, became the 15th state of the USA.

7 February 1795: The 11th Amendment to the Constitution was enacted to limit the jurisdiction of federal courts.

1 June 1796: Tennessee, formerly part of North Carolina, became the 16th state of the USA.

November 1796: The first presidential election contested by more than one candidate took place. Adams defeated Jefferson (who sought election as a Democratic Republican) by an electoral-college vote of 71–68.

1797: John Adams took the oath of office as the second President of the USA.

1800: Napoleon Bonaparte, who had come to power in France in 1799 in a *coup d'état* and would go on to become Emperor Napoleon I in 1804–15, secured the return of the Louisiana territory from Spain, in another secret treaty. This vast territory formalized the presence of revolutionary France in the centre of the North American continent.

1800: The newly created, purpose-built city of Washington, in its own District of Columbia so as not to be sited in any one state, became the third—and permanent—federal capital.

1800: Jefferson and his opponent, Aaron Burr, a resident of New York, and also a Democratic Republican, each won 73 votes in the Electoral College. The election was decided, according to the provisions of the Constitution, by the House of Representatives, who chose Jefferson (Burr became Vice-President).

1803: The Supreme Court established the power of judicial review.

1 March 1803: Ohio became the 17th state of the USA.

May 1803: President Jefferson dispatched future President James Monroe to Paris, where he successfully negotiated with Bonaparte the purchase of all French possessions on the North American mainland, for the sum of $15 m.; this territory, known as the Louisiana Purchase, covered more than 2m. sq km, and its acquisition doubled the USA's land area.

27 July 1804: The 12th Amendment to the Constitution was enacted, modifying the method by which the electoral college elects the President and Vice-President, so as to avoid a repeat of the difficulties encountered in the 1800 contest.

November 1804: President Jefferson stood for re-election against Charles Pinckney, a Federalist from South Carolina, and won an overwhelming victory in the Electoral College.

November 1808: James Madison, a Democratic Republican from Virginia, won a convincing victory over Pinckney in the presidential election; he was inaugurated in March 1809.

1812–14: The conflict known as the War of 1812 took place, in which the USA fought against the United Kingdom. British forces burned the Capitol Building and the White House in Washington, destroying precious documents, and great naval battles took place in the Atlantic, on Lake Champlain and on the Great Lakes. The USA, which sought to expel the British from Canada and thus the North American continent, failed in its objective, although it secured the demilitarization of the Great Lakes in the Treaty of Ghent, which concluded the War.

30 April 1812: Louisiana became the 18th state of the USA.

November 1812: President Madison stood for re-election and defeated DeWitt Clinton, a Federalist from New York.

1815: A US fleet was sent to combat piracy on the Barbary Coast of North Africa; this marked the new country's first significant involvement in international affairs.

1816: Congress granted a charter to the American Colonization Society, for the purpose of resettling American former slaves in Africa.

1816: Monroe, a Democratic Republican from Virginia, soundly defeated Rufus King, a Federalist from the Maine region of Massachusetts), in the presidential election.

11 December 1816: Indiana became the 19th state of the USA.

March 1817: Monroe took office as the fifth President of the USA.

10 December 1817: Mississippi became the 20th state of the USA.

1818: The Convention of 1818 partially ended disagreement between the USA and the United Kingdom over the so-called 'north-west boundary' by defining the border between the USA and the Canadian provinces as a straight line running along the 49th parallel (49°N) between the Great Lakes and the Rocky Mountains. Disagreement continued, however, as to where the border should lie west of the Rockies.

3 December 1818: Illinois became the 21st state of the USA.

1819: Spain ceded Florida to the USA under the terms of the Adams-Onís Treaty.

14 December 1819: Alabama became the 22nd state of the USA.

3 March 1820: Congress enacted the so-called Missouri Compromise legislation, which permitted slavery in Missouri but prohibited it in all other areas west of the Mississippi River and north of a line extending westward from the southern boundary of Missouri.

15 March 1820: Maine was separated from Massachusetts and became the 23rd state of the USA.

November 1820: President Monroe, who had proved highly popular, stood for re-election against John Quincy Adams of Massachusetts, also a Democratic Republican, and son of John Adams, the second President. Monroe won, receiving 231 of the 232 electoral-college votes.

10 August 1821: Missouri became the 24th state of the USA.

January 1822: Jehudi Ashmun led the first American former slaves to the West African coast, where they founded a colony. They would name this site Providence Island, and the colony would be called Liberia, meaning 'free land'.

25 April 1822: The Liberian settlers moved to a site on the African mainland, where they founded Monrovia, named after President Monroe.

2 December 1823: In the seventh State of the Union address of his administration, President Monroe outlined to the Congress the principles and policies that would later become known as the Monroe Doctrine. The doctrine opposed intervention by European powers in the affairs of any part of the American continent.

November 1824: John Quincy Adams stood again for the Presidency, this time as a National Republican. His three opponents were all Democratic Republicans: Andrew Jackson of South Carolina, Henry Clay of Kentucky and William Harris Crawford of Georgia. Although Jackson polled 155,872 votes, compared with Adams's 105,321, he failed to win the required Electoral-College majority. The House of Representatives sat to determine a winner and elected Adams.

1825: A treaty between the Russian and British Empires established the land border between Alaska, then a Russian possession, and the Northwest Territories of British Canada. The border between Alaska and the Canadian province of British Columbia would not be settled until 1903.

March 1825: John Quincy Adams took office as the sixth President of the USA, the first man to follow his father to the post.

26 October 1825: The Erie Canal opened to traffic when the first boat left Buffalo, on Lake Erie, and travelled down the network to the Mohawk River and the Hudson River to New York City.

4 July 1827: New York state abolished slavery.

1828: The part-Cherokee inventor and publisher, Sequoya, also known as George Geist or Guess, began publishing the first newspaper for indigenous Americans, written both in English and in the Cherokee language, for which he had devised an alphabet.

1828: The Democratic-Republican Party became the Democratic Party, and its candidate, Andrew Jackson, won the presidential election, defeating the incumbent President John Quincy Adams by a wide margin.

1829: Andrew Jackson took office as the seventh President of the USA.

November 1832: President Jackson stood for re-election and defeated Clay.

1 November 1835: Texan delegates convened a congress and unilaterally declared Texas's independence from Mexico. They formed a provisional Government, with the Virginia-born Texan Samuel Houston as President.

23 February–6 March 1836: A Mexican force of 4000 men commanded by General Antonio López de Santa Anna besieged 187 Texans at the Alamo, a fortified former Franciscan mission in San Antonio, Texas; none of the Texans survived.

21 April 1836: In the Mexican War, Houston led 800 Texans to victory against 3,000 Mexicans at the Battle of San Jacinto. Santa Anna surrendered and recognized Texan independence, although the Mexican parliament refused to honour his treaty.

15 June 1836: Arkansas became the 25th state of the USA.

November 1836: Martin Van Buren stood for the Presidency and soundly defeated his Whig opponents, the strongest of whom was the Virginian William Henry Harrison.

26 January 1837: Michigan became the 26th state of the USA.

March 1837: Martin Van Buren became the eighth President of the USA, and the first to have been born in the country after its independence.

1838: US troops carried out a policy formulated under the Jackson administration and declared unconstitutional by the US Supreme Court, but now nevertheless executed during Van Buren's presidency; they began evicting large numbers of indigenous Americans from their lands and forcibly marching them to Oklahoma, which had been designated 'Indian Territory'. The Cherokee of western North Carolina, who by now had established an elective self-government as the Cherokee Nation, were particularly harshly treated—at least one-fifth of the estimated 20,000 Cherokee who left their homelands died *en route*. Those who managed to complete the march found themselves in an arid area, where the agricultural skills they had developed over generations in the green hills of their homeland were of little use.

4 July 1840: President Van Buren signed into law the Independent Treasury Act, which separated the government's finances from those of the nation's banks, businesses and private citizens.

November 1840: President Van Buren sought re-election but was defeated by Harrison, who was inaugurated in March 1841. John Tyler, also a Virginian, became Vice-President.

4 April 1841: After just 31 days in office, President Harrison died of pneumonia. Vice-President Tyler became the first President to accede to the office on the death of his predecessor, as stipulated in the Constitution, and he became the 10th President. He completed the remainder of Harrison's term but chose not to stand for a term of his own.

1842: The first wagon train completed the 3,200-km (2,000-mile) journey along the Oregon Trail, linking Independence, in the state of Missouri, to the Columbia River in the Oregon territory. This and other such trails were the conduits through which American pioneers from the east settled the west.

1842: During the Tyler administration, government troops forcibly removed the Seminole people from their lands in southern Florida and marched them to Oklahoma.

1842: The Webster-Ashburton Treaty between the USA and Canada fixed the national border between the two countries from the eastern end of the border agreed by the Convention of 1818 to the Atlantic Ocean.

November 1844: James Knox Polk, a Democrat from North Carolina, defeated the Whig candidate, Clay, in the presidential election. He was inaugurated as the 11th President in March 1845.

3 March 1845: Florida became the 27th state of the USA.

29 December 1845: Texas, having voted in favour of annexation to the USA, became the 28th state of the Union.

1846: The Oregon Treaty permanently settled the remainder of the US–Canadian border, extending the 49th-parallel border westward to the Pacific (although granting Vancouver Island, which the parallel traversed, in its entirety to the Canadian territories).

May 1846: Mexico, which maintained its territorial claim to Texas and objected to the USA's annexation of it, disputed the boundary between Texas and the rest of Mexico. The USA regarded the Rio Grande (Río Bravo) as the boundary, whereas Mexico considered the more northerly Nueces River as the border. When Gen. Zachary P. Taylor and his US troops crossed the Nueces, Mexico interpreted his action as an invasion of Mexico and joined battle against them. The USA, considering that Gen. Taylor's men remained in US territory, declared that the Mexican troops, by engaging them in battle inside the USA, had committed an act of war.

13 May 1846: Congress declared war on Mexico.

28 December 1846: Iowa became the 29th state of the USA.

24 January 1848: As workmen were constructing a new sawmill in California, one of them discovered deposits of gold. Despite attempts to keep the discovery a secret, it soon became public knowledge, prompting mass immigration in to California (which remained under Mexican nominal control, although it had appeared likely that the USA would annex it after its ascendancy in the Mexican War).

2 February 1848: The Treaty of Guadalupe Hidalgo formalized the end of the Mexican War. Mexico ceded vast territories, to the USA, including not only Texas and California (already widely settled from the USA) but also large parts of present-day Arizona, Colorado, Nevada, New Mexico, Oklahoma, Utah and Wyoming.

29 May 1848: Wisconsin became the 30th state of the USA.

November 1848: Gen. Taylor stood for President as a Whig candidate. He narrowly defeated Lewis Cass, a Democrat representing Ohio. Van Buren, now representing the newly founded, anti-slavery Free Soil Party, was also defeated.

1849: The California 'Gold Rush' brought several hundred thousand 'forty-niners' from the 30 US states to the west coast in search of fortune. They arrived using a variety of routes, some obviating the rigorous land journey or the journey around South America by sailing to the isthmus of Panama, travelling overland across the isthmus from Atlantic to Pacific and then boarding Pacific steamships for the remainder of the journey (this method became known as the 'Vanderbilt Crossing', after the shipping line used). By the 1860 US Census, San Francisco, which in 1847 had had fewer than 1,000 inhabitants, was the 10th-largest city in the USA.

March 1849: Taylor was inaugurated as the 12th President of the USA. His Vice-President was Millard Fillmore, a Whig from New York.

9 July 1850: President Taylor died of cholera. Vice-President Fillmore succeeded him and became the 13th President.

9 September 1850: California became the 31st state of the USA, with slavery prohibited; New Mexico and Utah were made territories without a decision on slavery, under the so-called Compromise of 1850. The debate concerning slavery had polarized political opinion; many of the northern states having abolished the practice while the southern states argued for its retention. Some southern politicians advocated secession from the Union rather than acceptance of the abolition of slavery.

November 1852: In the presidential election (in which Fillimore had decided not to seek election to his own presidential term), Gen. Franklin Pierce, a Democrat from New Hampshire, defeated Lt-Gen. Winfield Scott, a Whig from Virginia. The two candidates had served together with distinction during the Mexican War.

March 1853: Pierce became the 14th President, of the USA. William Rufus King was inaugurated as Vice-President, but died the day after the ceremony; his office was not formally filled.

30 December 1853: The USA signed a treaty with Mexico whereby the it bought some 76,000 sq km (about 29,000 sq miles) of land in what is now southern New Mexico and Arizona. This purchase resulted in a clear definition of this part of the US–Mexican border.

1854: Representatives of the Whigs, the Free Soil Party and the right-wing so-called 'Know-Nothing Party', augmented by some defecting Northern Democrats dissatisfied with the Democratic Party's willingness to perpetuate slavery, grouped together to form the Republican Party.

November 1856: In the presidential election, James C. Buchanan, a Democrat from Pennsylvania, defeated the adventurer and former US Army officer, Maj. John Charles Frémont, a Republican from Georgia.

1857: Buchanan became the 15th President of the USA; John C. Breckenridge became Vice-President.

11 May 1858: Minnesota became the 32nd state of the USA.

14 February 1859: Oregon became the 33rd state of the USA.

16 October 1859: John Brown led a group of 18 anti-slavery activists in the seizure and occupation of the US Arsenal at Harper's Ferry, then in Virginia, later in West Virginia. The following day, US Marines under the command of Col (later Gen.) Robert Edward Lee regained control of the facility, killed 10 of Brown's men including two of his own sons and arrested Brown. The authorities charged Brown with treason, tried him, convicted him and sentenced him to death.

2 December 1859: John Brown was executed at Charlestown, then in Virginia, now the capital of West Virginia. His death strengthened the resolve of anti-slavery sentiment throughout the country—even in the South, where an increasing number of whites opposed both slavery and secession.

1860: Abraham Lincoln, born in Kentucky but long resident in Illinois, became the first modern Republican candidate to win the Presidency. He defeated three opposing candidates: the incumbent Vice-President, Breckinridge, a Democrat from Kentucky; John Bell, the Constitutional Union Party candidate from Tennessee; and Stephen Arnold Douglas, a Democrat, also representing Illinois.

20 December 1860: South Carolina voted to secede from the union, in defence of its right to formulate its own policies on matters such as slavery.

9 January 1861: Mississippi became the second state to secede. During the remainder of January, Florida, Alabama, Georgia, Louisiana and Texas also seceded, while Kansas was affiliated as the 34th state.

4 February 1861: The seven secessionist states sent delegates to a convention in Montgomery, Alabama, at which they adopted a provisional Constitution for the 'Confederate States of America'.

11 February 1861: The Mississippian Jefferson Davis was inaugurated as provisional President of the Confederacy.

March 1861: Abraham Lincoln took the oath of office as the 16th President of the USA. Hannibal Hamlin was inaugurated as Vice-President.

11 March 1861: The Confederacy ratified a permanent Constitution.

12 April 1861: US forces attempted to deliver provisions and supplies to the US Army post of Fort Sumter, at the entrance to the harbour at Charleston, South Carolina. Confederate forces opened fire on the Union installations. The confrontation marked the start of the conflict known as the American Civil War.

15 April 1861: President Lincoln ordered Union reinforcements to Charleston to regain control of the situation and protect the fort. In response, four more states (Virginia, Arkansas, Tennessee and North Carolina) seceded by the end of May. None of the secessions were recognized by the Union, which declared the states to be 'in rebellion'.

24 May 1861: The capital of the Confederacy was moved from Montgomery, Alabama to Richmond, Virginia.

21 July 1861: The Confederacy won the First Battle of Bull Run in Virginia.

6 November 1861: A presidential election took place in the Confederacy, at which Jefferson Davis was elected to serve a six-year term.

February 1862: Jefferson Davis was inaugurated as the first (and only) President of the Confederate States of America.

30 August 1862: The Confederacy won a second battle at Bull Run.

17 September 1862: Union forces led by Gen. George Brinton McClellan defeated a Confederate army led by Gen. Lee at Antictam, Maryland.

1863: Union forces under Gen. Ulysses Simpson Grant gained control the entire length of the Mississippi River.

1 January 1863: President Lincoln issued the Emancipation Proclamation, declaring that all slaves in any state, even a state now in rebellion against the Union, would henceforth be free.

20 June 1863: The western part of Virginia, which preferred to remain loyal to the Union, itself seceded from the secessionist state of Virginia and became West Virginia, the 35th state of the USA.

1–3 July 1863: The Battle of Gettysburg in resulted in a decisive victory for Union forces, led by Gen. George Gordon Meade, over the Confederate troops commanded by Gen. Lee.

31 October 1864: Nevada became the 36th state of the USA.

November 1864: President Lincoln stood for re-election and won a comfortable victory over his opponent, the Gen. George McClellan, who stood as a Democrat and advocated a peaceful resolution of the Civil War. Lincoln's new Vice-President was Andrew Johnson, a Tennessean former Democrat who had changed allegiance to the Republicans and fiercely opposed his state's secession from the Union.

15 November 1864: In Georgia, Gen. William Tecumseh Sherman led some 60,000 Union troops from Atlanta to the sea at Savannah, burning everything in his path and isolating Richmond from the western states of the Confederacy.

9 April 1865: Gen. Lee, with his Confederate troops surrounded on all sides by Union forces, surrendered to Gen. Grant at Appomattox, a small town in south-central Virginia.

14 April 1865: With the Civil War now virtually over, President Lincoln attended a play at Ford's Theatre in Washington, DC. During the performance, John Wilkes Booth shot Lincoln in the head. The President died of his injuries the following day.

15 April 1865: On President Lincoln's death in office by assassination, Vice-President Johnson succeeded and became the 17th President.

May 1865: The last outposts of Confederate resistance surrendered, ending the American Civil War.

6 December 1865: The 13th Amendment to the Constitution was enacted, abolishing slavery throughout the USA.

1867: President Johnson's Secretary of State, William Henry Seward, arranged to purchase Alaska from the Russian Empire for $7.2 m. He was ridiculed at the time by his critics who described the territory as 'Seward's Folly' or 'Seward's icebox'.

1 March 1867: Nebraska became the 37th state of the USA.

1868: President Johnson became the first US President to be impeached. He was charged on 11 counts of violating federal laws and of defying the authority of Congress. He was acquitted of all the charges and remained in office until the end of his term, choosing not to seek re-election.

9 July 1868: The 14th Amendment to the Constitution was enacted, protecting US citizens against abridgement of their citizenship rights.

November 1868: Gen. Grant stood for the presidency as a Republican candidate. He won a clear victory over Horatio Seymour, a Democrat from New York.

1869: The Union Pacific Railroad, constructed westward from Omaha, Nebraska, and the Central Pacific Railroad, constructed eastward from Sacramento, California, met at Promontory, in the Utah territory, establishing a rail link across the American continent.

1869: The territory of Wyoming granted women the right to vote, more than 50 years before the 19th Amendment to the US Constitution extended this right to women in the remainder of the country.

March 1869: Ulysses S. Grant took office and became the 18th President of the USA; Schuyler Colfax became Vice-President.

3 February 1870: The 15th Amendment to the Constitution was enacted, guaranteeing male citizens the right to vote regardless of 'race, colour or previous condition of servitude'.

1872: Congress voted to restore full civil rights to all citizens of the 11 Southern secessionist states, with the exception of 500 Confederate leaders.

1872: President Grant stood for re-election and defeated Horace Greeley of the Liberal Republican Party. Henry Wilson succeeded Colfax as Vice-President.

25 June 1876: A force of 2,500–4,000 men, mainly Sioux and Cheyenne, defeated 655 men of a regiment of the Seventh US Cavalry under the command of Lt-Col George Armstrong Custer at the Battle of the Little Big Horn, on the Little Big Horn River in what is now Montana. The event, often called 'Custer's Last Stand', began when Custer, having underestimated the Sioux troop strength, launched the first attack in an attempt to expand the USA's territory to the north-west.

1 August 1876: In the year marking the Centenary of the USA, Colorado became the 38th state of the Union.

November 1876: Rutherford Birchard Hayes, a Republican from Ohio, and Samuel Jones Tilden, the Democratic candidate from New York contested the presidential election. Tilden won the popular vote by 4,284,757 to Hayes's

4,033,950 but failed by one vote to win the required majority in the Electoral College. The Republicans contested election returns in Florida, Louisiana, Oregon and South Carolina, with the result that no clear winner had emerged.

29 January 1877: Congress enacted legislation to create an Electoral Commission, charged with settling the disputed presidential election—Hayes was eventually declared elected by a margin of one vote.

March 1877: Hayes was sworn in as the 19th President of the USA; William Wheeler was inaugurated as Vice-President.

November 1880: In the presidential election, James Abram Garfield, a Republican from Ohio, narrowly defeated Winfield Scott Hancock, a Democrat from Pennsylvania. Garfield was inaugurated as the 20th President in March 1881, with Chester Alan Arthur, a Republican from New York, as Vice-President.

2 July 1881: Charles Jules Guiteau shot President Garfield in Washington, DC; Garfield died on 19 September, Vice-President Arthur succeeding and becoming the 21st President.

November 1884: In the presidential election, Grover Cleveland, a Democrat from New York, defeated James Gillespie Blaine, a Republican from Maine.

March 1885: Grover Cleveland was inaugurated as the 22nd President, with Adlai Stevenson as his Vice-President.

November 1888: President Cleveland stood for re-election, but was defeated by Benjamin Harrison, a Republican born in Ohio but resident in Indiana, and a grandson of the 9th President, William Henry Harrison. He took office in March 1889, with Levi Morton as Vice-President.

2 November 1889: North Dakota and South Dakota both entered the Union, becoming the 39th and 40th states, respectively.

8 November 1889: Montana became the 41st state of the USA.

11 November 1889: Washington became the 42nd state of the USA.

3 July 1890: Idaho became the 43rd state of the Union.

10 July 1890: Wyoming became the 44th state.

15 December 1890: In South Dakota, the US Army captured and, allegedly during an escape attempt, killed the great Sioux Chief Katanka Yotanka, more commonly known as Sitting Bull. The event proved the impetus for the last major battle ever fought between indigenous Americans and US Cavalry troops.

29 December 1890: The Battle of Wounded Knee took place near Wounded Knee Creek in South Dakota; like the Battle of the Little Bighorn 14 years earlier, it was fought between the Seventh US Cavalry and the Sioux. At Wounded Knee, the Seventh Cavalry had detained Sioux Chief Big Foot and others on grounds that they may have been planning an uprising in retaliation for Chief Sitting Bull's death. The Cavalry claimed that a shot was fired by someone in the camp, giving them justification for their next action, which was to open fire directly into the camp.

November 1892: In the presidential election, the incumbent, Harrison, sought re-election but was defeated by former President Cleveland.

17 January 1893: In Hawaii, US factions were instrumental in the overthrow of the last Hawaiian monarch, Queen Liliuokalani. They established a republican form of government, led by the US-educated Sanford Ballard Dole.

March 1893: Cleveland for the second time took the oath of office as President of the USA; he is the only person ever to have served two non-consecutive terms, as both the 22nd and the 24th Presidents. Adlai Stevenson was Vice-President for his second term.

1896: The Supreme Court approved racial segregation with the proviso that such segregation could not prevent any group's development relative to that of another, the so-called 'separate but equal' doctrine.

4 January 1896: Utah became the 45th state of the USA.

November 1896: In the presidential election, William McKinley, a Republican from Ohio, defeated William Jennings Bryan, a Democrat representing Nebraska.

March 1897: McKinley took office as the 25th President of the USA, with Garrett Hobart as Vice-President.

December 1897: During Cuba's struggle for independence from Spain, the US, at this stage not involved in the dispute, sent the battleship *Maine* to Havana to protect the interests of US citizens there.

15 February 1898: The *Maine* was destroyed by an explosion in Havana harbour, with the loss of 260 US servicemen. (It was assumed at the time that Spanish forces had sunk the ship, although forensic studies conducted in 1969 found that a defective boiler on board had been the cause). The event precipitated US intervention in the conflict.

April 1898: Spain and the USA declared war on each other; the conflict became known as the Spanish-American War.

1 May 1898: A US Navy force under the command of Cdre (later Rear-Adm.) George Dewey, having been in Hong Kong when war was declared, arrived in the Spanish-held Philippines Islands, which Spain had occupied since 1571. There Dewey's force destroyed the entirety of the Spanish fleet stationed in Manila Bay.

18 July 1898: Spain admitted defeat and requested a negotiated settlement.

12 August 1898: The USA annexed the Hawaiian Islands.

13 August 1898: Rear-Adm. Dewey and the US Army officer, Gen. Wesley Merritt, captured Manila and effectively occupied the Philippines.

10 December 1898: The Treaty of Paris formalized the end of the Spanish-American War. Under the terms of the treaty, Spain recognized the independence of Cuba and ceded the Philippines, Puerto Rico and Guam to the USA.

November 1900: President McKinley won re-election, again defeating Bryan, this time by a slightly larger majority. His 'running-mate' (candidate for Vice-President) was Theodore Roosevelt, a Republican from New York.

March 1901: President McKinley was sworn in for his second term as President, and Theodore Roosevelt took the oath of office as Vice-President.

6 September 1901: Leon Czolgosz, an anarchist activist from Detroit, Michigan, shot President McKinley at the Temple of Music pavilion of the Pan American Exhibition in Buffalo, New York; at first it was thought that the President's injuries were not life-threatening.

14 September 1901: President McKinley died of his injuries in hospital in Buffalo. Vice-President Roosevelt, who was mountain-climbing at the time, succeeded and was sworn in as the 26th President of the USA.

29 October 1901: Czolgosz, having been tried, convicted and sentenced to death, was executed.

1903: The border between British Columbia and the US territory of Alaska was fixed by tribunal, allowing a strip of Alaskan territory to extend some 900 km southward along the Pacific coastline.

3 November 1903: Panama unilaterally declared its independence from Colombia. The USA immediately granted official diplomatic recognition of the independent nation of Panama; the construction of a canal across the Isthmus of Panama had been discussed in the US since the 'Vanderbilt crossings' of the Californian 'Gold Rush' in the mid-19th century.

18 November 1903: The USA and Panama signed the Hay-Bunau-Varilla Treaty, named after the US Secretary of State, John Milton Hay and the French engineer Philippe-Jean Bunau-Varilla. The latter, (a pupil of the creator of the Suez Canal), had devoted put decades of effort into realizing a canal across the Isthmus of Panama. The treaty committed the USA to guaranteeing the independence of Panama, thereby deterring Colombia from taking any action to recover the secessionist state. The treaty also granted the USA a 'perpetual' lease of the Panama Canal Zone, a strip of land extending 5 miles (8 km) on each side of the designated canal route. The USA agreed to pay Panama $10 million immediately and thereafter $250,000 per year; the latter figure would subsequently be increased in later treaties.

November 1904: President Roosevelt sought re-election against Alton Brooks Parker, a Democrat from New York, and won by a comfortable margin. He was inaugurated in March 1905, with Charles Fairbanks as his Vice-President.

1905: The US Isthmian Canal Commission assumed responsibility for the Panama Canal project, and Congress gave permission for the US Army Corps of Engineers to carry out the construction work, forecast to take 10 years.

1906: President Roosevelt won the Nobel Peace Prize for his mediation in the Russo-Japanese War.

16 November 1907: Oklahoma became the 46th state of the USA.

November 1908: In a presidential election, The Ohio-born Republican, William Howard Taft, defeated the Democrat nominee, Bryan (who was seeking election to the office for the third time).

March 1909: Taft took office as the 27th President; James Sherman was his Vice-President.

6 January 1912: New Mexico became the 47th state of the USA.

14 February 1912: Arizona became the 48th state of the USA, thereby bringing statehood to the entirety of the country's contiguous continental territory.

November 1912: President Taft stood for re-election, although his attempt to secure a second term was damaged by the announcement that Roosevelt intended to stand. When Taft defeated Roosevelt in the contest for the Republican nomination, Roosevelt established a new party, the Progressive Party, popularly known as the 'Bull Moose Party', to seek election to the presidency as a third-party candidate. In the presidential election itself, Roosevelt's presence divided the Republican vote, enabling the Democratic nominee, Thomas Woodrow Wilson, the Governor of New Jersey to win a majority in the Electoral College. The Progressive Party was formally dissolved five years later.

3 February 1913: The 16th Amendment to the Constitution granted the federal government the power to levy income taxes.

March 1913: Wilson took office as the 28th President of the USA and Thomas Marshall as Vice-President.

8 April 1913: The 17th Amendment to the Constitution allowed the election of US Senators—two from each state, serving six-year terms—by direct popular vote. State legislatures had previously elected Senators.

4 August 1914: Upon the outbreak of the First World War in Europe, President Wilson proclaimed US neutrality.

15 August 1914: The Panama Canal opened, a year ahead of schedule, although its cost had greatly exceeded its budget.

1916: The sparsely populated Rocky Mountain state of Montana, which had a single seat in the US House of Representatives, elected the first-ever woman member of the US Congress, Jeannette Rankin.

4 August 1916: The USA purchased the Danish West Indies (now the US Virgin Islands) from Denmark for $25m.

November 1916: President Wilson stood for re-election and narrowly defeated the Republican nominee, the former Governor of New York, Charles Evans Hughes.

6 April 1917: Congress declared war on Germany and the USA entered the First World War. The US joined the side of the Allied Powers against the Central Powers.

18 May 1917: Congress voted in favour of conscription, requiring all males aged 18–30 years to register for the draft.

26 June 1917: The first contingent of the American Expeditionary Force landed in France, under the command of Gen. John Joseph Pershing. Approximately 4m. US Army troops, three-quarters of them conscripted, and just under 600,000 US Navy personnel served in the First World War—some 117,000 were killed, more than one-half of them as a result of a pandemic of influenza.

8 January 1918: President Wilson submitted his peace plan, known as the 'Fourteen Points', to a joint session of Congress. One of the proposals was for a 'general association of nations' which would culminate in the founding, two years later, of the League of Nations.

11 November 1918: The First World War ended with the signing of the Armistice agreement in the Compiègne Forest in north-central France.

1919: For his services at the Paris Peace Conference, which formalized the end of the First World War, and his League of Nations proposal, President Wilson was awarded the Nobel Peace Prize.

16 January 1919: The 18th Amendment to the Constitution prohibited the manufacture, sale and transportation of intoxicating liquors; widely referred to as 'prohibition'.

19 November 1919: Congress voted against ratifying the Treaty of Versailles (which emerged from the Paris Peace Conference) and maintained its opposition to US membership of the League of Nations.

10 January 1920: The League of Nations was founded, although the refusal of Congress to assent to US membership damaged the organization's power.

26 August 1920: Following a protracted campaign for legislative change, the 19th Amendment to the Constitution granted women the right to vote.

November 1920: In the presidential election, Warren Gamaliel Harding, a Republican Senator from Ohio, defeated the Democrat, James Middleton Cox, the Governor of Ohio.

1921: Congress enacted an immigration quota system, whereby the number of people of a particular nationality allowed to enter the US in any one year could not exceed 3% of the number of that same nationality who had been resident in the USA in 1910.

March 1921: Harding took office as the 29th President of the USA. His running-mate, the Republican Governor of Massachusetts, John Calvin Coolidge, was sworn in as Vice-President.

2 July 1921: Congress ratified a separate peace treaty with Germany.

1922: The Secretary of the Interior, Albert Bacon Fall, was accused of receiving payments in return for granting secret leases to private-sector petroleum companies for exploration concessions in California and Wyoming, in what became known as the Teapot Dome scandal, after the name of the Wyoming concession. Fall was later convicted of corruption and served a one-year prison sentence.

2 August 1923: President Harding died in San Francisco, ostensibly of pneumonia. Coolidge succeeded him and became the 30th President of the USA.

1924: Congress finally granted full US citizenship to all indigenous Americans.

9 November 1924: The incumbent President Coolidge sought a presidential term of his own and won a comfortable victory over his two opponents, the Democratic nominee, John William Davis of West Virginia, and a third-party candidate, Robert Marion La Follette of Wisconsin, who had left the Republican Party to found the League for Progressive Political Action.

9 November 1924: Texas and Wyoming elected women as Governors (both were wives of former Governors).

March 1925: Coolidge was inaugurated as President, with Charles Dawes as Vice-President.

November 1928: The Republican Party chose Herbert Clark Hoover as its presidential nominee. The Democrats nominated the Governor of New York, Alfred Emanuel Smith, a Roman Catholic. Hoover won comfortably, owing to a national increase in anti-Catholic sentiment which Hoover himself did nothing to cultivate.

March 1929: Hoover began his term as the 31st President; Charles Curtis being inaugurated as Vice-President.

29 October 1929: The failure of a number of financial institutions, among other reasons, caused a loss of confidence in financial markets; investors sold equities in an attempt to gain liquid assets, The resultant sharp decline in stock values became known variously as the Stock Market Crash, the Wall Street Crash or Black Tuesday—by the end of the year, the capitalization value of US companies had declined by some $15,000m. The Crash began a period of economic hardship known as the Great Depression. Banks failed, causing frightened depositors at other banks to rush to withdraw their funds, thereby causing more banks to fail. Companies went bankrupt, and unemployment increased dramatically. The Great Depression continued to afflict the global economy throughout the 1930s.

November 1932: President Hoover stood for re-election, despite the rapid decline in his popularity during the Great Depression. Norman Mattoon Thomas, making the second of his six attempts to gain election as the nominee of the American Socialist Party, polled more votes in this election than in any other, although this remained less that 3% of the popular vote, and he secured none of the Electoral-College votes. The Democratic nominee was the Governor of New York, Franklin Delano Roosevelt, a cousin of the late 26th President, Theodore Roosevelt. Franklin Roosevelt won comfortably, after making electoral commitments to economic reform.

23 January 1933: The 20th Amendment to the Constitution moved the beginning date of a presidential term of office forward from March to 20 January, in order to shorten the period of time that an outgoing head of state can remain in office after having been defeated in a November election.

March 1933: Roosevelt took office as the 32nd President of the USA, with John Garner as Vice-President. His 'New Deal' reforms, designed to create jobs for unemployed people and to revive economic activity was approved amid opposition from elements within Congress.

5 December 1933: The 21st Amendment to the Constitution repealed the 18th Amendment; the manufacture, sale and transportation of intoxicating liquors was no longer a federal crime. The control of the illicit alcohol trade during the period of prohibition had been dominated by a number of organized-crime syndicates, and their power and influence had increased throughout society.

November 1936: President Roosevelt, whose 'New Deal' had proved highly popular, stood for re-election and won a convincing victory over the Republican nominee, the Governor of Kansas, Alfred Mossman Landon.

3 September 1939: France and the United Kingdom declared war on Germany following that country's invasion of Poland; the Second World War began. President Roosevelt declared US neutrality.

November 1940: President Roosevelt broke a long-standing precedent by seeking a third term as President. Despite vociferous criticism he again won by a wide margin—the defeated Republican opponent was Wendell Lewis Willkie.

January 1941: President Roosevelt began his third term and his ninth year in office. Iowa-born Democrat Henry Agard Wallace was sworn in as Vice-President in succession to Garner, who had repeatedly criticized President Roosevelt during his two terms of office, accusing him of excessive liberalism, and had unsuccessfully sought the Democratic nomination for the presidency.

7 December 1941: Japan (allied with Germany and Italy in the Second World War) launched a surprise attack on the 86-vessel US Pacific Fleet as it lay at anchor at Pearl Harbour, Honolulu, Oahu, Hawaii. About 100 Japanese aircraft scrambled from aircraft carriers dropped bombs while small submarines fired torpedoes; their combined efforts sank or disabled 18 US Navy vessels and destroyed 200 US military aircraft within a few hours.

8 December 1941: Japan carried out a similar attack on the US military base of Clark Field, near Manila, in the Philippines.

8 December 1941: Congress declared war on Japan. Three days later it declared war on Germany and Italy, and the country's entry into the Second World War was formalized.

1942: An estimated 110,000 US citizens of Japanese heritage were forcibly moved them into detention camps, where they would remain incarcerated for three years. US citizens of German or Italian heritage, of which there were many more, were not interned.

4–7 June 1942: The USA inflicted the first major defeat on Japan in the Battle of Midway, a remote coral atoll more than 2,000 km north-west of Honolulu.

8 November 1942: Lt-Gen. Dwight David Eisenhower led a joint British-US force (estimated to have numbered 290,000, roughly equally divided) in the invasion of occupied French North Africa.

8 September 1943: Italy surrendered shortly after Maj.-Gen George Smith Patton, Jr, led the US Seventh Army and the British-Canadian Eighth Army in a successful invasion of the Italian peninsula from North Africa.

6 June 1944: From the southern United Kingdom, US, British, Canadian, Free French and other Allied forces successfully crossed the English Channel and invaded France's Normandy coast during the 'D-Day' landings.

20 October 1944: Gen. Douglas MacArthur commanded a force of some 193,000 personnel and 738 ships in an attempt to liberate the Japanese-occupied Philippines.

November 1944: President Roosevelt won, again by a significant margin, an unprecedented fourth consecutive presidential election. His defeated opponent on this occasion was Republican Thomas Edmund Dewey, the Michigan-born Governor of New York.

January 1945: President Roosevelt began his fourth term and his 13th year as US President. Harry S Truman, the Democratic US Senator from Missouri, was sworn in as the new Vice-President.

3 February 1945: Gen. MacArthur's forces captured Manila; control of Luzon Island, the largest and politically dominant island of the Philippine group, was assured.

19 February 1945: In one of the bloodiest battles of the war, US Marines captured the Japanese air base on Iwo Jima, in the Volcano Islands group of the western Pacific Ocean.

1 April 1945: US forces landed on the island of Okinawa, in the Ryukyu group of the western Pacific, where the Japanese had a large air base. It would take the US 82 days of bloody combat before they finally won control.

12 April 1945: President Roosevelt died of a cerebral haemorrhage at Warm Springs, Georgia, a spa where he had often gone for his deteriorating health. Vice-President Truman succeeded him and became the 33rd President.

7 May 1945: After some factions of the German army had already begun surrendering on 4 May 1945, Germany surrendered unconditionally.

8 May 1945: Victory in Europe ('V-E') Day marked the end of the Second World War's European theatre. The Pacific theatre appeared further from resolution, however.

26 June 1945: A total of 50 charter member-states, meeting in San Francisco, California, USA, signed the United Nations Charter, creating the UN, the successor organization to the League of Nations.

16 July 1945: The world's first atomic bomb was detonated, as a test, in the desert near Alamogordo, New Mexico.

6 August 1945: The US dropped the first atomic bomb ever used in warfare on the Japanese city of Hiroshima; out of its population of about 350,000, estimates of the number killed either by the bomb itself or subsequently from the radiation vary between 80,000 and 220,000.

9 August 1945: The US dropped the second atomic bomb used in warfare on the Japanese city of Nagasaki; out of its population of about 250,000, estimates of the number killed by the bomb or its radiation vary between 40,000 and 75,000.

14 August 1945: Japan surrendered.

2 September 1945: Japan signed the official surrender document aboard the US battleship *Missouri* in Tokyo Bay; this date was designated Victory over Japan ('V-J') Day. The Second World War ended.

4 July 1946: The USA granted full independence to the Philippines.

1947: The granting of aid to Greece and Turkey, to combat Communist insurgency, occasioned the announcement of the so-called Truman Doctrine, which became central to US foreign policy.

1947: The US extended large amounts of aid to nine war-damaged European countries for rebuilding and recovery, under the European Reconstruction Program. This programme was widely known as the 'Marshall Plan', after its creator, the US Secretary of State, George Catlett Marshall who, in 1953 won the Nobel Peace Prize for this achievement.

30 April 1948: At the Ninth Inter-American Conference in Santafé de Bogotá (Colombia), 21 American nations, including the USA, agreed to form the Organization of American States (OAS).

August 1948: Following USA- and UN-sponsored elections, the Republic of Korea was created.

September 1948: Communists founded their own Democratic People's Republic of Korea in the north of the peninsula.

November 1948: President Truman sought re-election and narrowly defeated Dewey. He was inaugurated in March 1949, with Alben Barkley as his Vice-President.

29 June 1949: US troops withdrew from Korea, which was partitioned along the 38th parallel (38°N). The Communist Democratic People's Republic of Korea (North Korea) and the Republic of Korea (South Korea) were thus divided.

24 August 1949: The USA, Canada and 10 Western-European allies formed the North Atlantic Treaty Organization (NATO), in which the member-states committed themselves to defend one another.

1950: President Truman authorized the US production of the hydrogen bomb.

25 June 1950: North Korean troops crossed the 38th parallel, thereby invading South Korea; the Korean War began. North Korea, with the support of the People's Republic of China and the USSR, fought against South Korea, with the support of the USA and a UN-assembled coalition of 15 other member-states.

27 February 1951: The 22nd Amendment to the Constitution was enacted, formally limiting a President to two full elected terms of office. It also stipulated that a President who has served more than two years of a term to which a predecessor has been elected should be limited to one full elected term.

November 1952: Under the terms of the 22nd Amendment, incumbent President Truman was not eligible to seek re-election because he had already served more than two years of the late President Franklin Roosevelt's last term in addition to a full term of his own. Gen. Eisenhower was recruited to stand as a Republican, against the Democratic nominee, Adlai Ewing Stevenson (the Governor of Illinois and grandson of the former Vice-President of the same name). Gen. Eisenhower won with a large majority.

20 January 1953: Gen. Eisenhower was inaugurated as the 34th President, with Richard Milhous Nixon as his Vice-President.

July 1953: The Korean War ended with a truce signed at Panmunjom, a village in the demilitarized zone.

1954: Construction began on the St Lawrence Seaway, a joint Canadian-US project to enable ocean-going vessels to reach the Great Lakes between May and November (it freezes during the North American winter).

22 April–17 June 1954: Senator Joseph Raymond McCarthy, a Republican from Wisconsin, presided over televised congressional hearings into alleged Communist influence in the US Army. He would go on to accuse many people in government, the military and the intellectual community of being Communists.

17 May 1954: The Supreme Court unanimously ruled that 'separate educational facilities are inherently unequal' and, therefore, that the 'separate but equal' doctrine was in violation of the 14th Amendment to the Constitution.

2 December 1954: The Senate voted to censure Senator McCarthy, claiming that his accusations of Communism, rarely supported by substantial evidence, dishonoured the Senate and amounted to an abuse of his powers.

12 February 1955: The USA began to provide military-adviser assistance to the South Vietnamese Army, to help it fight against the Communist regime in North Viet Nam.

November 1956: President Eisenhower stood for re-election. Again his opponent was Stevenson, and again Eisenhower won decisively.

24 September 1957: President Eisenhower ordered federal troops to Little Rock, Arkansas, to enforce a federal court order that nine black students must be allowed to attend the previously all-white Central High School.

1958: The USA and Canada together founded the North American Aerospace Defense Command (NORAD).

31 January 1958: Explorer 1, the first US satellite, was launched into space.

1959: The US authorized its military advisers in Viet Nam to fight with the units they were training.

3 January 1959: Alaska became the 49th state of the USA.

25 April 1959: The St Lawrence Seaway came into operation, accepting its first ocean-going vessels.

21 August 1959: Hawaii became the 50th state of the USA.

15 September 1959: The Soviet Premier and Communist Party First Secretary, Nikita Khrushchev, became the first leader of the USSR to visit the USA.

November 1960: The incumbent Vice-President, Nixon, a Californian, stood for the presidency as the Republican nominee against Senator John Fitzgerald Kennedy, a Democrat from Massachusetts. In a close election, Kennedy won and, at the age of 43 years and five months, became the youngest person ever to win a presidential election. He was also the first Roman Catholic President.

20 January 1961: Kennedy took office as the 35th President of the USA. His running-mate, the Texan Democrat Lyndon Baines Johnson, became Vice-President.

29 March 1961: The 23rd Amendment to the Constitution was enacted, granting the right to vote in presidential elections to residents of the District of Columbia.

17 April 1961: Some 1,300 exiled Cubans residing in the USA, supplied by the USA and with the full knowledge of President Kennedy, attempted to invade Cuba and overthrow the Communist regime of Dr Fidel Castro Ruz, who had deposed a US-supported Government in 1959. They landed at the Bahía de los Cochinos (Bay of Pigs), on Cuba's south-western coast. Cuban troops promptly overcame them, killing 90 and taking the others prisoner.

19 March 1962: The Okinawa legislature in the US-occupied Ryukyu Islands was given greater autonomy.

22 October 1962: President Kennedy announced that aerial reconnaissance missions over Cuba on 16 October had revealed the presence of Soviet nuclear missiles in that country. The US Navy blockaded the island to prevent further deliveries of military equipment.

26 October 1962: Khrushchev stated that he would authorize the removal of the missiles from Cuba if the USA removed missiles from Turkey and undertook not to invade Cuba. President Kennedy replied that the Soviet missile programme in Cuba must be abandoned before any negotiation.

28 October 1962: The Cuban Missile Crisis eased when Khrushchev agreed to remove missiles from Cuba and not deliver any further missiles. US nuclear missiles were removed from Turkey in November.

25 July 1963: The Partial Nuclear Test Ban Treaty, signed by the USSR, the United Kingdom, the USA and the USSR, confined nuclear-explosives testing to underground sites.

22 November 1963: While on a campaign visit to Dallas, Texas, President Kennedy was assassinated by an assailant poised at a window of the Texas School Book Depository Building. Lee Harvey Oswald was accused of the crime but was himself assassinated before he could be brought to trial. Vice-President Johnson succeeded and was sworn in as the 36th President.

23 January 1964: The 24th Amendment to the Constitution was enacted, prohibiting the use of a poll tax or the requirement to pay any other tax as a means of denying any citizen the right to vote.

29 June 1964: The US Congress enacted the Civil Rights Act, which banned discrimination in places of public accommodation and in employment.

7 August 1964: The US Congress approved the Tonkin Resolution, which authorized military action in Viet Nam following reports of a North Vietnamese attack on two US destroyers.

November 1964: President Johnson stood for election to win a term of his own. His Republican opponent was Barry Morris Goldwater, a Senator from Arizona. Johnson won comfortably. He was inaugurated in January 1965, with Hubert H. Humphrey as his Vice-President.

February 1965: President Johnson authorized the bombing of suspected Communist positions in Viet Nam, a strategy he had condemned during the electoral campaign, claiming that Goldwater favoured its use.

6 August 1965: The US Congress enacted the Voting Rights Act, which banned the use of tests and other devices to deny the right to vote to certain minority groups.

3 October 1965: The national-origins quota system for immigration was abolished.

1 July 1966: Medicare and Medicaid, which provided government welfare support for health care for the elderly and the poor, were introduced.

1967: Domestic opposition to the undeclared war in Viet Nam increased, especially among young Americans, many of whom were liable to conscription under a lottery system based on the birthdays of all 19-year-old males.

10 February 1967: The 25th Amendment to the Constitution was enacted, specifying the procedure to be followed in case the office of President or Vice-President should become vacant through death, disability, resignation or removal.

1968: The US public's perception of the physical and financial costs of war in Viet Nam prompted President Johnson to propose cease-fire negotiations (at this time more than 500,000 US troops were stationed in Viet Nam).

3 April 1968: The Baptist clergyman, civil-rights activist and Nobel Peace Prize winner Rev. Martin Luther King, Jr, was assassinated in Memphis, Tennessee. White escaped prisoner James Earl Ray would be arrested, tried, convicted and sentenced to life in prison for the crime, but later even Rev. King's widow would demand his release on grounds that the evidence against him was inadequate.

5 June 1968: The Democratic Senator Robert Francis Kennedy, the younger brother of the late President Kennedy (in whose administration he had served as Attorney-General), was shot in a corridor of the Ambassador Hotel in Los Angeles, California, while campaigning for the Democratic Party presidential nomination. He died the following day. A Jordanian national, Sirhan Bishara Sirhan, was arrested, tried with his murder, convicted of the crime and sentenced to life imprisonment.

November 1968: The presidential election was contested by the incumbent Vice-President, Humphrey, as the Democratic nominee (Johnson having declined to seek re-election), Eisenhower's Vice-President, Nixon, as the Republican nominee, and the right-wing Governor of Alabama, George Corley Wallace. Nixon narrowly defeated Humphrey.

18 January 1969: Talks aimed at resolving the conflict in Viet Nam began in Paris (France).

20 January 1969: Richard Nixon took office as the 37th President, with Spiro Agnew as his Vice-President.

8 July 1969: President Nixon ordered the phased withdrawal of US troops from Viet Nam to begin.

1970–71: The US organized invasions into Cambodia and Laos to sever North Vietnamese supply lines.

1 July 1971: The 26th Amendment to the Constitution was enacted, lowering the voting age from 21 years to 18 years.

21 February 1972: President Nixon made an eight-day state visit to the People's Republic of China.

15 May 1972: The Ryukyu Islands were returned to Japan by the USA.

22 May 1972: President Nixon made an eight-day visit to the USSR, during which the two countries signed the Strategic Arms Limitation (SALT) Treaty.

17 June 1972: Five men were arrested following a break-in at the offices of the Democratic National Committee in the Watergate complex in Washington, DC. It was alleged that they were attempting to place a listening device. Investigations by the media gradually revealed evidence that the detained men were connected with President Nixon's re-election campaign, although any such links were denied.

August 1972: In Viet Nam, the USA had withdrawn all combat troops by this time but continued to supply military assistance to South Vietnam, notably air support. The protracted peace negotiations failed.

10 October 1972: The *Washington Post* alleged that the Federal Bureau of Investigations believed that the Watergate incident formed part of a systematic campaign of espionage on the part of President Nixon's campaign team.

November 1972: President Nixon stood for re-election, against the Democratic Senator from South Dakota, George Stanley McGovern. Nixon won by one of the widest margins in US history, McGovern emerging victorious in only one of the 50 states, Massachusetts.

20 January 1973: President Nixon began his second term of office.

27 January 1973: Conscription ended.

30 January 1973: Seven presidential aides were convicted of burglary and 'wire-tapping' (attempting to listen to others' telephone conversations using an unlawful listening device).

7 February 1973: The Senate voted to authorize the Senate Select Committee on Presidential Campaign Activities to investigate the Watergate incident. The Committee, chaired by Senator Sam Ervin and informally known as the Ervin Committee, began televised hearings on 17 May.

29 March 1973: The last US troops were withdrawn from Viet Nam.

30 April 1973: The Attorney-General and two senior presidential advisers resigned, another adviser was dismissed.

August 1973: It was announced that a federal attorney's office in Baltimore, Maryland, was investigating Vice-President Agnew for possible offences of tax fraud, extortion and bribery, arising from his days as Governor of Maryland.

7 August 1973: The Ervin Committee closed its first session, having established that recordings were made of all conversations in the Presidential offices during Nixon's tenure. The President refused to surrender the recordings to the investigation; they were eventually procured by means of legal action.

10 October 1973: Vice-President Agnew resigned.

12 October 1973: The provisions of the 25th Amendment were used for the first time. Gerald Rudolph Ford, the Minority Leader in the House of Representatives, became the first person ever to be appointed, rather than elected, to the office of Vice-President.

6 December 1973: Ford was inaugurated as Vice-President.

7 December 1973: An 18-minute silence in one of the recordings made in the presidential offices was discovered; the explanations provided by the presidential staff were widely considered to be insufficient.

9 May 1974: The House of Representatives' Judiciary Committee opened impeachment hearings against President Nixon.

27 July 1974: The House Judiciary Committee recommended that three articles of impeachment be brought against President Nixon.

9 August 1974: Rather than endure the impending impeachment proceedings, President Nixon resigned, becoming the first President in US history to do so. Vice-President Ford succeeded and was sworn in as the 38th President of the USA, becoming the first President to have been elected neither to the office of President nor to the office of Vice-President.

September 1974: President Ford controversially granted former President Nixon an unconditional pardon, with full immunity against any criminal proceedings.

January 1975: The North Vietnamese launched a renewed offensive; the remaining US personnel, who were mostly civilians, were evacuated.

April 1975: The South Vietnamese regime surrendered to the North Vietnamese.

November 1976: President Ford stood for election, hoping to win the office in his own right. The Democratic nominee, James Earl (Jimmy) Carter, Jr, the Democratic candidate, emerged as the narrow victor.

1977: Panama and the USA agreed to replace the 1903 Hay-Bunau-Villars Treaty with two new treaties which gave Panama sovereignty over the Panama Canal Zone in the same year and granted Panama sovereignty over the canal itself by December 1999. The USA, however, retained the right to defend the canal's neutrality indefinitely.

20 January 1977: Carter took the oath of office as the 39th President; Walter Mondale was inaugurated as Vice-President.

1 October 1979: The USA formally transferred the Panama Canal Zone to Panama.

4 November 1979: Islamist extremist students took 63 US citizens hostage in the US Embassy in Tehran (Iran).

4 January 1980: President Carter announced punitive measures against the USSR following its invasion of Afghanistan; among them were embargoes on the sale of grain and advanced-technology products.

November 1980: President Carter sought re-election but lost by a significant margin to the Republican candidate, the former film actor and incumbent Governor of California, Ronald Wilson Reagan.

20 January 1981: Following agreement on the return of frozen Iranian assets to the value of some $8,000 m., the US hostages were released on the day President Carter left office. An attempt to free the hostages by military means had failed in April 1980, when two aircraft collided, killing eight service personnel.

20 January 1981: Reagan took the oath of office as the 40th President of the USA. His Vice-President was George Herbert Bush, the former head of the Central Intelligence Agency.

November 1982: After nearly three years, President Reagan lifted the embargo on the sale to the USSR of technical equipment for the petroleum and natural-gas industries.

25 October 1983: US Marines, with some assistance from other Caribbean countries, carried out a full-scale invasion of the Caribbean island country of Grenada, for the purpose of ousting a Marxist regime.

November 1984: President Reagan stood for re-election and won a convincing victory over the Democratic candidate, former Vice-President Mondale. Mondale's running-mate, Geraldine Anne Ferraro, was the first woman ever to receive the nomination of a major political party as its vice-presidential candidate.

26 November 1984: Iraq and the USA re-established full diplomatic relations; Iraq had terminated relations in 1967, owing to US support for Israel during the Arab-Israeli conflict.

19–20 November 1985: President Reagan and the Soviet leader, Mikhail Sergeyevich Gorbachev, met for two days of private talks in Geneva.

14 April 1986: US forces bombed the Libyan cities of Tripoli and Benghazi, claiming the action was in retaliation for terrorist attacks by Libya.

November 1986: The 'Iran-Contra' affair emerged, involving secret military sales to Iran and the use of the resulting funds, illegally, for the support of right-wing guerrilla forces in Nicaragua.

8 December 1987: President Reagan met Gorbachev at a 'summit' meeting in Washington, DC, at which the two leaders signed the Intermediate Nuclear Forces Treaty, providing for the dismantling of all US and Soviet missiles with a range of less than 5,500 km but more than 500 km.

November 1988: The incumbent Vice-President, Bush, stood for the presidency as the Republican candidate, against the Democrat Michael Stanley Dukakis, the Governor of Massachusetts. Bush won comfortably.

20 January 1989: Bush took office as the 41st President, and James Danforth Quayle as Vice-President.

20 December 1989: Following increasing tensions, the USA invaded Panama, restoring the authority of democratic government and removing the regime of Gen. Manuel Antonio Noriega Morena. Noriega evaded capture, but was discovered and surrendered two weeks later.

1990: Members of NATO and the Warsaw Pact (the defence alliance of European Communist countries) signed the Treaty on Conventional Forces in Europe.

3 January 1990: The US took Noriega to the Florida to be tried on charges of drugs-trafficking, 'money-laundering' and racketeering. He was subsequently convicted and sentenced to 20 years' imprisonment.

August 1990: After Iraq invaded Kuwait, the US imposed economic sanctions against Iraq.

7 August 1990: In 'Operation Desert Shield', US forces embarked for Saudi Arabia, for the stated purpose of protecting that country and its neighbours from further invasion by Iraq.

1991: The USA and the USSR signed the Strategic Arms Reduction Treaty.

January 1991: In 'Operation Desert Storm', a US-led multi-national force drove occupying Iraqi forces out of Kuwait and restored the country's monarchy through aerial bombardment of targets in Iraq. Sanctions were subsequently maintained against Iraq and bombardment resumed on a number occasions during the 1990s and early 2000s.

December 1991: The dissolution of the USSR marked the end of the 'Cold War', the ideological opposition between the USA and its allies for one part and the USSR and its allies for the other, which had threatened to escalate into conflict, possibly involving nuclear weapons, since the conclusion of the Second World War.

1992: The USA agreed to provide financial assistance to the Russian Federation and other former Soviet republics, both for nuclear disarmament and for economic restructuring.

1992: The USA deployed some 28,000 troops to Somalia to restore order and facilitate the distribution of food supplies in the country. The troops were subsequently withdrawn after becoming increasingly involved in the conflict.

7 May 1992: The 27th Amendment to the Constitution was enacted, stipulating that any legislation providing for salary increases for members of Congress may not take effect until after the subsequent congressional election.

November 1992: President Bush sought re-election against the Democratic nominee, William Jefferson (Bill) Clinton, the Governor of Arkansas. The Independent candidate, Henry Ross Perot, was considered to have divided the conservative vote and enabled Clinton to win a majority in the Electoral College.

20 January 1993: Bill Clinton took the oath of office as the 42nd President; his Vice-President was Albert Gore, Jr.

26 February 1993: A bomb exploded under the tallest buildings in New York City, the two towers of the World Trade Center. Four men, believed to be linked to the wealthy Saudi-born Islamist dissident, Osama bin Laden, were convicted of involvement in the attack in March 1994.

1993: President Clinton successfully introduced measures to reduce the government's budget deficit, but failed to gain congressional approval of major health-care reforms.

1993: A 'Great Flood' of the Mississippi River basin covered some 8m. acres (3.2m. ha) and caused damage to an estimated value of $12,000m.

1993–94: President Clinton and his wife, Hilary Rodham Clinton, were the subject of numerous allegations of financial impropriety, relating to their participation in a savings bank and property-development company which had collapsed during the President's tenure as Governor of Arkansas in 1989. The Clintons denied all accusations of impropriety in what became known as the 'Whitewater' affair, although further allegations were subsequently made and became the subject of official investigations. Five senior presidential appointees subsequently resigned, following allegations of their involvement in the 'Whitewater' affair.

1994: Following economic sanctions against Haiti, that country's personal regime was removed and democratic government restored, with the assistance of a UN-approved, US-led military force.

1 January 1994: The North American Free Trade Agreement (NAFTA), negotiated with Canada and Mexico (and signed in December 1992) by President Bush and approved by Congress during the Clinton administration, took effect; the Agreement liberalized trade between the three countries of continental North America.

October 1994: The Secretary of Agriculture, Mike Espy, resigned in connection with allegations of improper financial dealings with Arkansas companies which had supported President Clinton.

November 1994: In legislative elections, the Republicans gained control of both houses of Congress for the first time since 1953.

1995: President Clinton used emergency powers to grant a $20,000m. loan to the USA's NAFTA partner, Mexico, during a severe financial crisis in that country. he subsequently proposed a 'Free Trade Area of the Americas' for 34 Western-Hemisphere nations, to enter into effect by 2006.

19 April 1995: The detonation of a bomb outside a federal office building in Oklahoma City caused 169 deaths; initial opinion that the attack had been executed by Islamist extremists was swiftly dismissed and concern centred on the activities of right-wing US militias. Timothy McVeigh, a member of one such group, was subsequently arrested in connection with the attack—he was convicted and sentenced to death in June 1997 and executed in June 2001.

November 1995: An *impasse* in budget negotiations between the executive and the legislature culminated in the temporary suspension of most federal-government operations.

November 1995: US-sponsored peace negotiations at Dayton, Ohio led to a settlement in the conflicts in the former Socialist Federal Republic of Yugoslavia, most notably of the civil war in Bosnia and Herzegovina, where US ground troops joined multi-national peacekeeping forces at the end of the year.

1996: Congress granted an increase in the national minimum wage in return for presidential agreement to certain reductions in welfare expenditure.

1996: Despite some progress in trade negotiations, there was tension in US relations with the People's Republic of China following Chinese military exercises in the Straits of Taiwan.

12 March 1996: Relations with Canada, Mexico and the countries of the European Union (EU) were impaired by the enactment of the 'Helms-Burton' proposals (named after the two Senators who proposed the legislation) officially known as the Cuban Liberty and Democratic Solidarity Act, which imposed a US boycott of all companies which maintained foreign investment in Cuba.

May 1996: President Clinton's successor as Governor of Arkansas resigned after being convicted of fraud in connection with the 'Whitewater' affair.

June 1996: A Senate committee alleged that Hilary Clinton and a number of presidential officials had obstructed a federal investigation into the 'Whitewater' affair.

27 July 1996: Concerns about acts of terrorism within the USA were heightened by a bomb at the Olympic Games in Atlanta, Georgia, and the subsequent unexplained explosion of a commercial aircraft shortly after its departure from New York, which killed 238 people.

November 1996: A strong economy ensured that President Clinton's attempt to secure re-election was successful, although there were later allegations of improper campaign financing by the Democratic Party. His unsuccessful Republican opponent was Robert Joseph Dole, who had resigned from his position in the Senate to contest the election. Perot stood again, with less success than in 1992.

January 1998: A former junior employee on the presidential staff, Monica Lewinsky, was required to give evidence in a civil action brought against President Clinton by Paula Jones, who claimed that he had sexually harassed her in 1991. The congressionally appointed special prosecutor investigating the 'Whitewater' affair, Kenneth Starr, was subsequently authorized by the Attorney-General to investigate whether a sexual relationship between Clinton and Lewinsky had taken place in 1995, as alleged by Lewinsky. President Clinton denied any improper conduct with Lewinsky and, following testimony to a Grand Jury on the matter, denied having made any attempt to influence any person, including Lewinsky, to give false testimony.

7 August 1998: Bombs exploded near the US embassies in Kenya and Tanzania. More than 250 people were killed in the attacks and several thousands were injured. Suspicion was directed at Islamist extremist groups, most notably al-Qa'ida (base), led by bin Laden. US forces attacked sites in Sudan and Afghanistan believed to be used by the organization.

17 August 1998: Following an offer of immunity from prosecution for perjury made by Starr to Lewinsky, President Clinton gave further evidence to the Grand Jury and publicly admitted that an inappropriate relationship had taken place with Lewinsky, although he denied obstructing the legal investigation. The President asked for the public's forgiveness. Republicans began to demand that the President resign or be impeached.

11 September 1998: Starr's report to Congress alleged that President Clinton had committed 11 offences of perjury and obstruction of justice, which constituted grounds for impeachment. No evidence of wrongdoing was presented in connection with 'Whitewater'.

8 October 1998: The House of Representatives voted to commence an impeachment inquiry against President Clinton.

8 December 1998: Following protracted legal arguments, the House Judiciary Committee voted to approve four articles of impeachment against President Clinton, for 'high crimes and misdemeanours' which had resulted in the President's violation of his constitutional duty.

16 December 1998: A further series of bombing raids was aimed at targets in Iraq.

18 December 1998: The House of Representatives approved two of the articles of impeachment (alleging, respectively, perjury before a Grand Jury and obstruction of justice), rejecting the other two. The case was, in accordance with the Constitution, referred to the Senate.

7 January 1999: The impeachment hearings began in the Senate, before the President of the Supreme Court, William Rehnquist.

12 February 1999: Following over one month of hearings and legal arguments, the Senate voted on the articles of impeachment against President Clinton (a two-thirds' majority was required for conviction). On the count of perjury, the President was acquitted by 55 votes to 45, and on the count of obstruction of justice 50 Senators voted to convict and 50 to acquit. Clinton thus remained in office.

March 1999: US forces participated in the NATO-led aerial bombardment of Yugoslavia, following the country's refusal to comply with international demands for an end to the perceived persecution of the ethnic-Albanian minority in the province of Kosovo and Metohija.

late June 2000: The Supreme Court ruled that Elián González, a six-year-old Cuban boy who had been in the USA since he was shipwrecked on a vessel carrying Cuban refugees to the country should be returned to Cuba. Despite opposition from Cuban groups within the USA and resistance from the relatives with whom the boy was residing, he was eventually returned to Cuba after the intervention of the US military.

7 November 2000: In the Presidential election, the incumbent Vice-President, Gore, was opposed by the Republican George Walker Bush (the Governor of Texas and son of the 41st President, George Herbert Bush) and seven other candidates. It became evident during the hours after the polls closed that possession of a majority in the 538-member Electoral College would depend on the 25 college mandates from Florida. In turn, the result in Florida was extremely close; exit polls having initially declared the state for Gore, then for Bush and then having abandoned predictions. The state was eventually declared for Bush, pending the result of an automated recount of the votes, giving him 271 Electoral-College mandates. However, amid numerous accusations of electoral irregularities and incorrectly registered votes, Gore disputed the result and subsequently began legal action. In concurrent legislative elections, the Democrats made a net gain of five senatorial seats (Hilary Clinton was elected to represent New York), equalizing the total number of Senators from each party at 50, although Republicans retained control of the House of Representatives.

late November 2000: Protracted legal arguments over the recounting of votes in Florida prevented the final declaration of that state's results. Following a number of state and federal court judgments, Gore's request for recounts of the votes in Florida by electoral staff was approved by that state's Supreme Court. However, the Republicans argued that no provision for any such 'manual' recounts had been made in the electoral legislation, and appealed to the US Supreme Court.

13 December 2000: The US Supreme Court upheld the Republican appeal against the 'manual' recounts, by a margin of five judges to four. With all legal recourses exhausted, Gore conceded defeat in Florida, and thereby nationally. Final results subsequently revealed that Gore had been successful in the national popular vote, winning 48.4% of valid votes cast, compared with 47.9% for Bush.

January 2001: President Clinton made strenuous attempts to accelerate the peace-processes in Northern Ireland and in the Middle East—during his presidency his participation in these processes had become as personal as it was official—the other participants approved of his final effort, fearing that President-elect Bush might seek to reduce US involvement in international affairs not directly related to US interests.

mid-January 2001: Less than one week before leaving office, President Clinton agreed to admit giving misleading testimony to an Arkansas court regarding his relationship with Lewinsky and to make a contribution towards the cost of the investigation into his alleged obstruction of justice in the matter. In return, the investigation was discontinued.

19 January 2001: A number of presidential pardons issued by Clinton on his last full day in office were widely criticized, including that given to Marc Rich (a wealthy financier who had fled the USA while under investigation on charges of tax-evasion in 1983 and whose wife had contributed to the Democratic campaign fund) and to Clinton's own half-brother, convicted of drugs-trafficking in 1985.

20 January 2001: George Walker Bush was inaugurated as the USA's 43rd President, with Richard Cheney as Vice-President. Bush's first cabinet included a number of individuals who had served in his father's administration, although one nominee resigned before appointment and a number of other nominees, most notably the Attorney-General-designate, John Ashcroft, were subjected to intensive questioning before gaining the required approval of the Senate.

late January 2001: The Bush administration immediately suspended the implementation of a number of environmental regulations endorsed by Clinton during his final weeks as President.

February 2001: President Bush's first budget included significant reductions in levels of personal taxation; a commitment to this policy had been an important feature of his electoral campaign.

March 2001: President Bush was subjected to international criticism after his announcement that the USA would not ratify the Kyoto Protocol on climate change, claiming the agreement to be detrimental to US interests. Further criticism was attracted by the President's support for National Missile Defence (NMD), a modification of the Strategic Defence Initiative (SDI) of the 1980s. Whereas SDI had forseen a network of space-based missiles capable of intercepting missiles launched towards the USA, NMD envisaged the systems being largely terrestrially based. The development of the proposals would require the amendment of a number of nuclear-arms-limitation treaties, an eventuality which proved unpopular with a number of countries, including the Russian Federation.

April 2001: A US Navy reconnaissance aircraft, claimed by the USA to have been on a mission in international airspace over the South China Sea, was involved in a collision with one of two Chinese fighter aircraft monitoring its movements. The damaged US aircraft was forced to make an emergency landing at a military airbase on the Chinese island of Hainan; it was subsequently impounded by the Chinese authorities and its crew, numbering 24, were detained. The Chinese aircraft and its pilot were lost. Both countries blamed the other's aircraft for causing the incident; the crew was released following 11 days of intensive negotiations, following a US statement expressing regret for the presumed death of the Chinese pilot and for entering Chinese airspace. The aircraft was dismantled and returned to the USA in June.

May 2001: Senator James Jeffords of Vermont left the Republican Party to sit in the Senate as an independent, thus giving the Democrats a majority in the chamber.

11 September 2001: Four commercial passenger aircraft were hijacked shortly after their departures on internal flights; two from Boston, one from New York and one from Washington, DC. The two aircraft originating in Boston, both bound for Los Angeles, were diverted to New York and each was flown into one of the towers of the World Trade Center, causing the towers to collapse. The aircraft originating in Washington, DC, bound for San Francisco, was returned to Washington, where it was directed into the buildings of the US Department of Defense, known as the Pentagon. All major public buildings were subsequently evacuated and US air-space closed. The fourth aircraft crashed in a rural area of Pennsylvania. The four aircraft were carrying a combined total of 266 passengers and crew, all of whom died. The precise total of fatalities at the World Trade Center remained unknown in November 2001, but was thought to exceed 5,000, including many firefighters and police-officers working at the scene when the towers collapsed. A total of 125 people were considered to have died at the Pentagon. President Bush stated that those responsible would be punished; following a number of claims of responsibility which were dismissed, it emerged that the authorities suspected the events to have been ordered by bin Laden and his al-Qa'ida network, (four men believed to be linked to the organization had been convicted of planting a bomb beneath the building in March 1993 and a further four were standing trial in the USA charged with involvement in the bombings of the US embassies in Kenya and Tanzania in August 1998 at the time of the attacks). Bin Laden, believed to be in Afghanistan, where he had been granted refuge by the extremist Islamist Taliban regime, subsequently rejected accusations that he had ordered the attacks, although he announced his approval of the events.

13 September 2001: President Bush announced that he intended that those responsible for the attacks should be captured, 'dead or alive'. Diplomatic activity began, with the aim of forming an international 'coalition against terrorism'. US military forces were sent to positions in South Asia.

8 October 2001: US forces, assisted by those of other countries (most notably the United Kingdom), began aerial bombardment of Afghanistan, demanding that the Taliban regime surrender bin Laden. A number of Islamic countries (and bin Laden himself) claimed that the bombardment amounted to an attack on Islam—the USA reiterated that this was not the intention, insisting that it was seeking to eradicate terrorism.

18 October 2001: The four men accused of involvement of the bombings of the US embassies in Kenya and Tanzania were convicted and sentenced to life imprisonment. All had confirmed links to al-Qa'ida.

12 November 2001: A passenger aircraft crashed shortly after leaving a New York airport bound for the Dominican Republic, killing all 260 people on board and a number (not considered to exceed six) on the ground. Initial investigations suggested that the incident had been an accident, although the possibility of sabotage was not excluded.

Uruguay

c. **2000 BC:** Tribal units of the Charrúa and the Tupí-Guaraní people emerged. Other indigenous groups included the Yaro, Chaná and Bohane people.

AD 1516: The Spanish, led by Juan Díaz de Soís, entered the Río de la Plata and discovered the region.

1603: Hernando Arias de Saavedra, the first Spanish governor of the Río de la Plata region, introduced the first cattle and horses.

1680: The Portuguese founded Colonia de Sacramento as a rival to Buenos Aires, which was located on the opposite shore of the Río de la Plata.

1726: Montevideo was founded by the Spanish as fortress against the Portuguese at Colonia de Sacramento. Thus Montevideo became the port and station of the Spanish fleet in the South Atlantic.

1807: A British force of 10,000 men captured Montevideo and occupied it for several months, after which it moved against Buenos Aires, Argentina, where it was defeated.

1808: Montevideo declared its independence from Buenos Aires.

18 May 1811: The army of José Gervasio Artigas, who had rebelled against Montevideo's military commander, won its most important victory against the Spanish in the Battle of Las Piedras.

1814: Forces from Argentina entered Montevideo.

January 1815: The Orientales (Uruguayans) defeated the Argentines at Guayabos and regained Montevideo.

1820: Forces from Brazil occupied Montevideo.

1825: Argentine and Brazilian forces clashed in Uruguay, in what became known as the Cisplatine War.

27 August 1828: With British mediation, Brazil and Argentina signed the Treaty of Montevideo, whereby both countries renounced their claims to the territories which then became the República Oriental del Uruguay.

18 July 1830: The Constitution of the República Oriental del Uruguay was officially approved.

1836: Political rivalry became violent and led to the formation of the first political parties, the Colorados and the Blancos, which eventually became the Partido Colorado (PC) and the Partido Nacional (PN).

1876–86: Uruguay was ruled by a military regime, under which the dominance of central authority in Montevideo was established over regional *caudillos*.

1903–07 and 1911–15: During the presidencies of the Colorado leader José Batlle y Ordóñez, progressive policies led to the creation of the first Latin American welfare state.

1904: Civil war fought between the Colorado Government and Blanco rebels ended in victory to the former.

1919: A new Constitution divided the executive branch of government between a President and eight other members of a National Council.

1929: José Batlle y Ordóñez died.

1933: The rule of the National Council of Government was overthrown in a *coup d'état* led by Gabriel Terra.

1942: The *batllistas* returned to power.

1951–66: The presidency was rotated, on an annual basis, among the nine members of the restored National Council of Government.

1959: The PN defeated the Colorados in a general election to take power for the first time in the 20th century.

27 November 1966: A new Constitution, reintroducing the presidential system of government, was approved by referendum; simultaneously, Gen. Oscar Daniel Gestido of the PC was elected President.

6 December 1967: Vice-President Jorge Pacheco Areco assumed the presidency following the death of Gen. Gestido.

1968–73: The activities of the Tupamaro urban guerrilla movement grew in prominence throughout the country.

28 November 1971: Juan María Bordaberry Arocena, a Colorado, won the presidential election.

1973: The Tupamaro urban guerrilla movement was suppressed by a sustained military offensive. Military intervention in civilian affairs led to the closure of Congress and its replacement by an appointed 25-member Council of State (subsequently increased to 35 members). The Partido Comunista and other left-wing groups were banned, as were political meetings.

27 July 1974: Army officers were placed in control of the major state-owned enterprises.

1976: Despite the Government's announcement that there would be a return to democracy, persecution of political figures continued, and the number of political prisoners held in 1976 was thought to have reached 6,000. The USA suspended military aid, accusing the regime of human-rights violations.

12 June 1976: President Bordaberry was deposed by the army because of his refusal to countenance any return to constitutional rule.

14 July 1976: The recently formed Council of the Nation elected Dr Aparicio Méndez Manfredini to the presidency.

1979: The USA decided to restore a fraction of its former military aid to Uruguay.

30 November 1980: A draft of a new constitution, under which the armed forces would continue to be involved in all matters of national security, was rejected in a referendum by 57.8% of voters.

1 September 1981: A retired army general, Gregorio Alvarez Armellino, was appointed by the Joint Council of the Armed Forces to serve as President during the transition period to full civilian government.

1983: The reluctance of the Government to permit greater public freedom and to improve observance of human rights caused serious social unrest.

August 1983: The Government suspended all public political activity and reserved the right to impose a new constitution without consultation.

September 1983: The Plenario Intersindical de Trabajadores—Convención Nacional de Trabajadores (PIT—CNT), the country's main trade-union federation, organized the first industrial action for 10 years. The action was supported by 500,000 workers.

July 1984: Talks between the Government, the Partido Colorado and the Unión Cívica (a Christian-democratic party) resumed, and the parties obtained several important concessions, including the right to engage in political activity.

3 August 1984: Political parties, with the exception of the Partido Nacional, agreed to the Government's proposals for a transitional set of laws and the formation of a National Security Council. All restrictions on political activity were withdrawn.

25 November 1984: In presidential and legislative elections the Partido Colorado, led by Dr Julio María Sanguinetti Cairolo, secured a narrow victory over the opposition. The Colorado Party won 38.6% of the ballot in the presidential contest, but did not gain an absolute majority of seats in Congress.

11 February 1985: The military regime relinquished power.

1 March 1985: President Sanguinetti was inaugurated, as was a Government of national unity, incorporating representatives of other parties.

14 March 1985: All political prisoners were released under an amnesty law.

1986–88: Numerous strikes were staged by public-sector employees, primarily against government economic policy.

22 December 1986: An amnesty law, the 'punto final', was approved, which brought to an end trials of military and police personnel accused of human-rights violations during the military dictatorship. The President was made responsible for any further investigations.

February 1987: A campaign was initiated by left-wing parties, trade unions and human-rights groups to organize a petition to force a referendum on the issue of the amnesty law.

16 April 1989: In a referendum, 52.6% of the votes were cast in favour of maintaining the amnesty law.

26 November 1989: In presidential elections, Luis Alberto Lacalle Herrera, of the Partido Nacional, won with 37% of the votes cast, while the Partido Colorado candidate, Jorge Batlle Ibáñez received 30%. In congressional elections the Partido Nacional gained the most seats but failed to obtain an overall majority.

1 March 1990: Lacalle was inaugurated as President and announced the conclusion of an agreement, the 'coincidencia nacional', between the two principal parties, whereby the Partido Colorado undertook to support proposed legislation on economic reform, in return for the appointment of four of its members to the Council of Ministers.

26 March 1991: The Treaty of Asunción was signed by Argentina, Brazil, Paraguay and Uruguay, committing the four countries to the establishment of a common market, to be known in Portuguese as the Mercado Comum do Sul and in Spanish as the Mercado Común del Sur (Mercosul/Mercosur–the Southern Common Market).

May 1991: Former President Sanguinetti, the leader of the Foro Batllista faction of the Partido Colorado, withdrew his support from the Government in protest against proposed privatization legislation, thus forcing the resignation of the Minister of Public Health, the faction's sole representative in the Council of Ministers.

September 1991: Congress narrowly approved privatization legislation.

1992–93: Industrial unrest in the public sector intensified.

February 1992: A new Council of Ministers was appointed.

October 1992: Approximately 30% of the electorate voted for a full referendum to be held on the partial amendment of the privatization legislation.

13 December 1992: At a referendum, 71.6% of voters supported the proposal for a partial repeal of the privatization legislation.

January 1993: President Lacalle reorganized the Council of Ministers, but did not reconstruct the coalition with the Partido Colorado.

9 March 1993: The Minister of Public Works and Transport resigned following the withdrawal of legislative support for the Government by two influential factions of the Partido Nacional.

27 November 1994: In presidential and legislative elections the Partido Colorado won a narrow victory, with 32.5% of the vote, compared to the Partido Nacional's 31.4% and 30.8% for the Encuentro Progresista—Frente Amplio (EP—FA), a predominately left-wing alliance. The leading presidential candidate of the Partido Colorado, Sanguinetti, was pronounced President-elect.

31 December 1994: Mercosul/Mercosur was established; its provisions came into effect the following day.

February 1995: After establishing a 'governability pact' with the Partido Nacional, which provided for a coalition government, President-elect Sanguinetti announced the appointment of a new Council of Ministers.

1 March 1995: Sanguinetti was inaugurated President.

8 December 1996: Amendments to the Constitution, which included the abolition of the system (Ley de Lemas) whereby each political party could present more than one presidential candidate, and the introduction of a provision for second-round presidential elections, were approved by 50.5% of voters in a referendum.

13 January 1997: The new amendments to the Constitution came into force.

26 January 1998: Minor changes were made to the composition of the Council of Ministers.

3 October 1998: Following the death of the Vice-President, Hugo Batalla, further changes were made to the Council of Ministers.

August 1999: The Minister of Livestock, Agriculture and Fishing resigned amid a crisis in the farming sector.

31 October 1999: In legislative elections the traditional dominance of the Partido Colorado and the Partido Nacional was ended when the EP—FA became the largest single party in both the Chamber of Deputies and the Senate, where it won 40 and 12 seats, respectively. In the first round of the presidential elections Tabaré Vázquez, of the Partido Nacional, gained 38.5% of the vote, compared with 31.3% for the Colorado candidate, Batlle.

28 November 1999: Batlle defeated Vázquez in the second round of presidential elections with 51.6% of the votes cast. He was aided by the support of the Partido Nacional, under a formal accord.

1 March 2000: Batlle was inaugurated President and appointed a new Council of Ministers.

August 2000: President Batlle announced the formation of a commission to investigate the fate of those who disappeared during the military dictatorship.

18 February 2001: An attempt by the EP—FA to force a referendum on recently passed privatization legislation failed, owing to a lack of public support.

April 2001: The Uruguayan authorities suspended exports of beef following the discovery of cattle infected with foot-and-mouth disease near the border with Argentina.

July 2001: The PIT—CNT led a general strike to protest against the continuing privatization programmes and high unemployment.

September 2001: The Government announced that the foot-and-mouth epidemic had ended. There had been a total of 2,057 outbreaks during the epidemic, and the decision to vaccinate the country's entire herd was hailed as a success.

Venezuela

***c.* AD 500:** The region was inhabited by indigenous peoples belonging to three principal groups—the Cariban, Arawak and Chibcha.

1498: On his third voyage to the Americas, Cristoforo Colombo, the Italian-born navigator employed by the Spanish crown (known in English as Christopher Columbus), was the first European to set foot on the soil of what is now Venezuela.

1499: The country was given its name Venezuela (meaning 'Little Venice') by the Spanish explorer Alonso de Ojeda.

1521: The first Spanish settlement on the mainland was established at Cumaná.

1528: King Carlos V of Spain mortgaged the whole colony to the German banking house of Welser. After two decades of searching for the mythical land of El Dorado, the enterprise collapsed and Venezuela returned to Spanish rule.

1567: The settlement of Caracas was founded and became the most prosperous city and province.

1620: The cultivation of coca was introduced and became one of the region's most lucrative export commodities, attracting Spanish immigrants to the colony.

1718: The Viceroyalty of New Granada (which governed present-day Venezuela, Colombia, Ecuador and Panama) was established to replace the Viceroyalty of Peru, which had previously administered the region.

1724–39: The Viceroyalty of New Granada was reincorporated into Peru.

1728: The Guipuzcoana Company was granted a monopoly of the colony's trade by the Spanish Government, in exchange for monitoring piracy and smuggling activity.

1739: The Viceroyalty of New Granada was restored.

1777: Venezuela became a Spanish Captaincy-General, with an enhanced degree of administrative autonomy from Santafé de Bogotá, the capital of New Granada.

1783: The Superintendency of Venezuela, which encompassed the approximate territory of the present-day state, was established.

1810: The *criollos* (of mixed European and indigenous descent) of Caracas deposed the Spanish governor and established a governing junta.

5 July 1811: The junta declared that Venezuela was independent and began devising a new constitution.

25 July 1812: The first independent republic collapsed as a result of increasing royalist and popular opposition.

1813: Simón Bolívar, a revolutionary *criollo*, returned to Venezuela from exile in Colombia. He was proclaimed Liberator of his country after his victorious army reached Caracas and declared a new republic.

September 1814: The second republic fell after Bolívar's army was defeated in Caracas.

1819: Under the leadership of Bolívar, New Granada achieved independence from Spain.

June 1821: Bolívar's troops fought the decisive Battle of Carabobo that liberated Caracas from Spanish rule.

August 1821: Delegates from Venezuela and Colombia met at the border town of Cúcuta to formally sign the Constitution of the Republic of Gran Colombia. Bolívar was named President of the Republic, which, from 1822, also encompassed the liberated Ecuador.

1826: A pro-independence rebellion was led by Gen. José Antonio Páez, but suppressed by forces loyal to Bolívar.

May 1830: Gran Colombia was dissolved and Venezuela became a separate republic.

1830–46: Under the presidency of Gen. José Antonio Páez, Venezuela was consolidated as a state, aided by a stable economy, which derived benefit from high prices for its principal export commodity, coffee.

1846: Gen. Páez was succeeded by Gen. José Tadeo Monagas, precipitating a period of dictatorial rule.

1854: Slavery was abolished.

1858–70: Gen. Monagas was ousted and an intermittent civil war followed as provincial leaders engaged in a power struggle with the central regime.

1870–88: Antonio Guzmán Blanco established a personalist regime and initially brought peace to the country. However, his popularity soon declined and, following a period of civil unrest, marked by rioting, Guzmán sought exile in Paris, France.

1892–98: Gen. Joaquín Crespo led a military regime.

1899–1908: Gen. Cipriano Castro assumed the presidency after Crespo was killed.

1908–35: Juan Vicente Gómez deposed Castro and assumed power. Democratic opposition to Gómez's repressive regime emerged from the middle-class student movement. The most prominent leaders included Rómulo Betancourt, who founded the centre-left Acción Democrática (AD) and Dr Rafael Caldera Rodríguez.

1936–41: Congress selected Gen. Eleazer López Contreras to serve as President.

1941: The AD achieved legal recognition.

1941: Congress selected Isaías Medina Angarita to replace López as President.

1945: Collaboration between AD and progressive elements in the military led to military-civilian *coup d'état* in which Medina was replaced by Rómulo Betancourt as head of a revolutionary junta.

March 1946: The junta decreed a new electoral law in which universal suffrage for all citizens over 18 years, including women, was enshrined, and all political parties were legalized. The centre-right Comité de Organización Política Electoral Independiente (COPEI) was subsequently founded by Caldera Rodríguez.

5 July 1947: A new Constitution was promulgated, formalizing many of the legislative changes made by the regime in the preceding years.

14 December 1947: The AD won overwhelming majorities in the presidential and congressional elections.

15 February 1948: Rómulo Gallegos of the AD was inaugurated as President.

24 November 1948: The military, led by Lt-Col Carlos Delgado Chalbaud, deposed Gallegos after he refused to share political authority with COPEI. The military junta abrogated the 1947 Constitution and proscribed the AD.

13 November 1950: Lt-Col Delgado was assassinated by Gen. Rafael Simón Urbina.

27 November 1950: Dr Germán Suárez Flammerich was appointed President of the military junta.

December 1952: Col (later Gen.) Marcos Pérez Jiménez seized power.

22 January 1958: Pérez was overthrown by a military junta, led by Adm. Wolfgang Larrazábal.

30 October 1958: The corruption and oppression of Gen. Pérez's regime led the previously antagonistic political parties to form the Pact of Punto Fijo. The Pact committed AD and COPEI to share positions in the state administration, to respect the outcome of democratic elections and to control their respective constituencies.

7 December 1958: The AD candidate, Rómulo Betancourt, won the presidential election with 47% of the votes cast. Adm. Larrazábal had resigned from the Navy the previous month to participate in the election.

February 1959: Betancourt was inaugurated as President. He was to prove the first constitutionally elected President to complete a full five-year term of office.

10–14 September 1960: The first conference of the Organization of the Petroleum Exporting Countries (OPEC) was held in Baghdad, Iraq, with representatives from Iran, Iraq, Kuwait, Saudi Arabia and Venezuela in attendance.

23 January 1961: A new Constitution was promulgated which guaranteed a wide range of civil liberties and created a bicameral legislature.

1962–63: A guerrilla group, the Fuerzas Armadas de Liberación Nacional (FALN) sabotaged petroleum pipelines and bombed the US embassy in Caracas.

1 December 1963: Dr Raúl Leoni, of the AD, was elected President with 32.8% of the votes cast. In concurrently-held legislative elections the AD won the largest number of seats in both Houses of Congress but failed to gain an overall majority in either. Leoni was inaugurated on 11 March 1964.

30 October 1966: Supporters of former President Gen. Pérez participated in an abortive military uprising.

1 December 1968: Caldera Rodríguez was elected President, receiving 29.1% of the votes cast. He was inaugurated in March 1969.

June 1970: The Port of Spain Protocol was signed in Trinidad by the foreign ministers of Venezuela and Guyana. Under the terms of the agreement, Venezuela agreed not to assert any claim to sovereignty over a large area of Guyana to the west of the Essequibo river, to which Venezuela had, since 1962, repeatedly laid claim.

9 December 1973: Carlos Andrés Pérez Rodríguez of the AD was elected President, receiving 48.6% of the votes cast in a presidential election. In concurrent congressional elections the AD attained absolute majorities in both Houses of Congress. Pérez assumed office on 11 March 1974.

3 December 1978: The presidential election was won by COPEI leader, Dr Luis Herrera Campíns, who received 46.6% of the votes cast. The candidates from the two main political parties secured almost 90% of the total presidential ballot—the AD candidate, Luís Piñerúa Ordaz, won 43.3%. In congressional elections COPEI and AD won an equal number of seats in the Chamber of Deputies and in the Senate. Campíns was inaugurated in March 1979.

1981: A deteriorating economic situation provoked social unrest and a succession of guerrilla attacks.

4 December 1983: Dr Jaime Lusinchi of the AD was elected President with 56.8% of the votes cast. The AD also won a majority of seats in the Congress. Lusinchi officially assumed power on 2 February 1984.

29 October 1988: A total of 14 Venezuelan fishermen, suspected of being Colombian drugs-smugglers, were killed by units of the Venezuelan army at Guafitas, near the town of El Amparo. The massacre provoked protests and rioting erupted in several cities. A group of 20 soldiers who were arrested in connection with the killings were subsequently released on the ruling of a military court.

4 December 1988: The former President, Carlos Andrés Pérez Rodríguez, was re-elected with 52.9% of the votes cast. The COPEI candidate received 40.4%. In congressional elections the AD lost its overall majority.

February 1989: Following Pérez's inauguration at the beginning of the month, a series of adjustments were implemented, designed to halt Venezuela's economic decline. These measures, which included increases in the prices of petrol and public transport, provoked rioting throughout the country. The Government introduced a curfew and suspended various constitutional rights in order to quell the disturbances; official sources claimed that some 246 people had died during the protests.

March 1989: The curfew was revoked, and all constitutional rights were restored, after wages had been increased and the prices of some basic goods were 'frozen'.

28 March 1989: An agreement was reached with Colombia on the establishment of a border commission to negotiate a settlement concerning a territorial dispute over the maritime boundaries in the Gulf of Venezuela.

December 1989: The first direct elections for state governorships were characterized by a high degree of abstention and significant gains by left-wing parties, which had campaigned energetically against corruption allegedly rife in the main political parties.

March 1990: The Presidents of Venezuela and Colombia signed the 'San Pedro Alejandro' document, by which they pledged to implement the border commission's proposals.

20 November 1990: Clashes between demonstrators and the security forces, during protests against the Government's economic policies, resulted in several fatalities.

January 1992: President Pérez gave recognition to Colombia's claims to territorial rights in the Gulf of Venezuela.

4 February 1992: An attempt to overthrow the President by rebel army units, identified as members of the Movimiento Revolucionario Bolivariano 200 (MRB-200), was defeated by armed forces loyal to the Government. Simultaneous rebel action in the cities of Macay, Maracaíbo and Valencia ended when the leader of the *coup d'état*, Lt-Col Hugo Chávez Frías, broadcast an appeal for their surrender. A number of constitutional guarantees were immediately suspended, and press and television censorship was imposed to exclude coverage of Chávez, who had received considerable passive popular support.

5 March 1992: President Pérez announced a series of political and economic reforms, including the introduction of legislation for immediate reform of the Constitution. In addition, increases in the price of petrol and electricity were suspended and price controls on a number of basic foodstuffs and on medicine were introduced.

10 March 1992: Two members of COPEI were appointed to the Council of Ministers in an effort to broaden the base of support for the Government.

9 April 1992: The Government approved the restoration of constitutional guarantees, suspended since the attempted *coup d'état* in February.

11 June 1992: Following the withdrawal of the COPEI members from the Council of Ministers, President Pérez effected a minor cabinet reshuffle.

27 November 1992: An attempt by rebel members of the armed forces to overthrow President Pérez was suppressed by forces loyal to the Government. The insurrection was reported to have been instigated by members of MRB-200. The Government introduced a state of emergency and suspended the Constitution.

21 May 1993: The Senate voted to endorse a Supreme Court ruling that sufficient evidence existed for President Pérez to be brought to trial on corruption charges. The charges concerned allegations that Pérez and two former government ministers had, in 1989, embezzled US $17m. from a secret government fund. Pérez was subsequently suspended from office and, in accordance with the terms of the Constitution, replaced by the President of the Senate, Octavio Lepage, pending the election by Congress of an interim President.

5 June 1993: Congress elected Ramón José Velásquez, an independent Senator, as interim President. He subsequently completed a comprehensive reorganization of the Council of Ministers.

31 August 1993: Congress voted in favour of the permanent suspension from office of Pérez, regardless of the legal outcome against him.

5 December 1993: Caldera Rodríguez, the candidate of a newly-formed party, the Convergencia Nacional (CN), was elected President, having obtained 30.5% of the votes cast in the presidential election. In concurrently-held legislative elections the AD won a combined 71 seats in the two chambers; the COPEI won 68 while CN and its electoral ally the Movimiento al Socialismo (MAS) secured 60 seats.

2 February 1994: President Caldera took office for a further term and installed a Council of Ministers, which included new portfolios with responsibility for economic reform, youth affairs, higher education and science and technology.

February 1994: The AD and COPEI established a legislative pact using their combined majority representation in Congress in order to gain control of the major legislative committees. In response, President Caldera warned that he would dissolve the legislature and organize elections to a constituent assembly should the government programme encounter obstructions in Congress.

27 June 1994: The Government announced that, in view of the deepening economic crisis and the virtual collapse of the banking sector, it was to assume extraordinary powers. Six articles of the Constitution were suspended concerning guarantees including the freedom of movement and freedom from arbitrary arrest. Price controls and a single fixed exchange rate were also introduced in order to address the problems of a rapidly depreciating currency and depleted foreign exchange reserves.

7 November 1994: The Government announced the deployment of the National Guard to patrol the major cities in response to rising levels of crime.

15–16 March 1995: The arrest of some 500 alleged subversives (including activists of the MRB-200) by the security forces prompted demonstrations in Caracas in support of demands for the release of those detained and in opposition to the militarization of Venezuela's cities.

6 July 1995: With the promulgation of legislation allowing for government intervention in the economy without recourse to further emergency measures, the six suspended constitutional guarantees were restored, except in designated areas along the borders with Colombia and Brazil where criminal activity persisted.

14 March 1996: President Caldera announced a major reorganization of the Council of Ministers.

February 1997: Following a succession of violent incidents on the countries' joint border, Venezuela and Colombia signed an accord creating a bilateral verification commission to conduct investigations into border incidents.

21 March 1997: As part of a reorganization of cabinet portfolios, a new post, that of Minister of State for Social Security, was created to oversee the reform of the social security system.

29 July 1997: The leader of the attempted *coup d'état* of February 1992, Lt-Col (retd) Hugo Chávez Frias, founded his own political party, Movimiento V República (MVR).

30 June 1998: The Minister of Finance, Freddy Rojas Parra, resigned in the midst of an escalating economic crisis.

8 November 1998: Legislative elections to the Chamber of Deputies were won by the Polo Patriótico, an alliance of small, mainly left-wing parties led by Chávez's MVR, with 34% of the votes cast. The AD secured 21.7% of the ballot while CN won only 2.4%. In the Senate the AD won 19 seats compared with 12 seats won by the MVR.

6 December 1998: Chávez was elected President, with 56.2% of the votes cast, despite an attempt by the AD and COPEI to preserve their long-standing political dominance by uniting behind a single candidate, Henrique Salas Römer, who received the support of some 40% of voters.

2 February 1999: Chávez was inaugurated and a new Council of Ministers was installed.

22 February 1999: An ambitious social improvement plan, the Proyecto Bolívar 2000, was launched to rehabilitate public property and land through voluntary civil and military action. Such pledges prompted a series of rural and urban public land occupations—none of which were forcibly ended.

25 April 1999: A national referendum, organized to ascertain levels of support for the convening of a constituent assembly to draft a new constitution, demonstrated 81.5% support for such a body.

25 July 1999: Elections to the Constituent Assembly resulted in an outright victory for Polo Patriótico, which secured 120 of the 131 seats. The leadership of COPEI and AD subsequently resigned.

12 August 1999: The Constituent Assembly declared a national emergency in the public institutions to allow radical reforms to be implemented. Congress was subsequently put into recess indefinitely.

9 September 1999: The Constituent Assembly and Congress signed an agreement of political coexistence that guaranteed the full functioning of the legislative branch from 2 October until the promulgation of a new constitution.

15–18 December 1999: Continuous torrential rains caused floods and mudslides along the coast around La Guiara (the principal port serving Caracas) in Vargas state. An estimated 30,000 people were killed and 200,000 left homeless.

15 December 1999: The draft constitution was approved by 71.2% of the votes cast in a referendum. The Constitution renamed the country the Bolivarian Republic of Venezuela, extended the President's term from five to six years and replaced the bicameral Congress with the unicameral Asamblea Nacional (National Assembly).

30 December 1999: The new Constitution was promulgated.

24 January 2000: President Chávez appointed Isaias Rodríguez to the new position of executive Vice-President.

31 January 2000: The Constituent Assembly was dissolved after it formally delegated its powers to a newly created 21-member legislative commission. The legislative commission was designated as an interim body until the formation of the National Assembly.

15 February 2000: The 'Guasdualito Declaration' was signed by Venezuela and Colombia, which included a commitment to joint border area development work, and co-operation on bilateral issues.

25 May 2000: Three days before voting was scheduled to begin in presidential, legislative, gubernatorial and local elections, the Supreme Tribunal of Justice (which had recently replaced the Supreme Court under the terms of the new Constitution) suspended the elections, citing technical faults with the electronic voting system.

30 July 2000: Chávez won 59.7% of the valid votes cast in the postponed presidential election, compared with the 37.5% of the ballot polled by Francisco Arias Cárdenas, the Governor of Zulia, and the 2.7% secured by Claudio Fermín, the former Mayor of Caracas. In elections to the new National Assembly, the MVR secured 93 seats, the AD 32, Proyecto Venezuela eight and the MAS six, while the once powerful COPEI only won five seats.

16 August 2000: Chávez was inaugurated as President for a six-year term.

October 2000: President Chávez renewed Venezuela's claim to the Essequibo region in Guyana.

7 November 2000: The National Assembly approved an 'enabling law', allowing the President to decree, without parliamentary debate, legislation in a number of policy areas, including public finance and land reform. The decree legislation would be valid for a period of one year from promulgation.

25 November 2000: Colombia briefly recalled its ambassador to Venezuela after a representative of the guerrilla group, the Fuerzas Armadas Revolucionarias de Colombia, was allowed to speak in the Venezuelan National Assembly.

December 2000–January 2001: There were a series of cabinet reshuffles which included the appointment of Alvaro Silva Calderón as Minister of Energy and Mines to replace Alí Rodríguez Araque (who had been appointed Secretary-General of OPEC).

3 December 2000: Despite resistance from trade unions and the International Labour Organization, a referendum was held on whether all trade-union leaders should be suspended until new elections for their positions could be held.

President Chávez accused the trade-union leaders of corruption and undemocratic behaviour while his critics accused him of attempting to eradicate a powerful opposition faction. The proposal was approved by 65% of the participating electorate.

4 February 2001: A minor cabinet reshuffle was implemented in which José Vicente Rangel became the first civilian to be appointed Minister of National Defence. A further minor rearrangement of portfolios took place in March.

1 November 2001: The USA recalled its Ambassador to Venezuela in response to comments made by President Chávez against the US bombing campaign in Afghanistan.